13th International Conference on New Developments and Applications in Optical Radiometry (NEWRAD 2017)

Journal of Physics: Conference Series Volume 972

Tokyo, Japan
13 - 16 June 2017

Editor:

Julian Grobner

ISBN: 978-1-5108-5932-6
ISSN: 1742-6588

Printed from e-media with permission by:

Curran Associates, Inc.
57 Morehouse Lane
Red Hook, NY 12571

Some format issues inherent in the e-media version may also appear in this print version.

Printed by Curran Associates, Inc. (2018)

For permission requests, please contact the Institute of Physics at the address below.

Institute of Physics
Dirac House, Temple Back
Bristol BS1 6BE UK

Phone: 44 1 17 929 7481
Fax: 44 1 17 920 0979

techtracking@iop.org

Additional copies of this publication are available from:

Curran Associates, Inc.
57 Morehouse Lane
Red Hook, NY 12571 USA
Phone: 845-758-0400
Fax: 845-758-2633
Email: curran@proceedings.com
Web: www.proceedings.com

Table of contents

Volume 972

13th International Conference on New Developments and Applications in Optical Radiometry

13–16 June 2017, Tokyo, Japan

Accepted papers received: 16 February 2018
Published online: 2 March 2018

Preface

Papers

Earth Observation

Detector-based radiometry: scale realizations

Other topics

IOP Conf. Series: Journal of Physics: Conf. Series **972** (2018) 011001 doi:10.1088/1742-6596/972/1/011001

Julian Gröbner, Guest editor and chairman of the NEWRAD scientific committee
Physikalisch-Meteorologisches Observatorium Davos, World Radiation Center (PMOD/WRC)
Dorfstrasse 33
7260 Davos Dorf, Switzerland

Abstract
The 13th International Conference on New Developments and Applications in Optical Radiometry (NEWRAD 2017) was held at the scenic waterfront district Odaiba, in Tokyo Japan, from 13 to 16 June 2017. The local organising committee and its chairman Tatsuya Zama from the National Metrology Institute of Japan organised an outstanding conference with Oral and Poster sessions, alongside memorable events such as the scenic Yakatabune cruise through Tokyo Harbour. The conference was attended by 119 participants, the majority coming from National Metrology Institutes and the remote sensing community.

The Topics of the conference were grouped in the following categories:

- Earth observation (EAO)
- Solar/stellar radiometry (SSR)
- Quantum optics technologies (QOT)
- Source-based radiometry (SBR)
- Detector-based radiometry: scale realizations (DBS)
- Detector-based radiometry: applications (DBA)
- Optical properties of materials/components (OPM)
- Other topic (OT)

129 presentations were submitted, of which 44 were selected for oral presentations, while the remaining were presented as posters during two extended lunch-time breaks, giving the chance to the participants for in-depth discussions.

The next NEWRAD conference will be organised by John Lehman, from the National Institute for Standards and Technology (NIST) at Boulder, Colorado, U.S. in early summer 2020. For additional information concerning NEWRAD2020 please contact either the chair of the local organising committee, John Lehman, or the chair of the NEWRAD scientific committee, Julian Gröbner.

IOP Conf. Series: Journal of Physics: Conf. Series **972** (2018) 011002 doi:10.1088/1742-6596/972/1/011002

Peer review statement

All papers published in this volume of *Journal of Physics: Conference Series* have been peer reviewed through processes administered by the proceedings Editors. Reviews were conducted by expert referees to the professional and scientific standards expected of a proceedings journal published by IOP Publishing.

IOP Conf. Series: Journal of Physics: Conf. Series **972** (2018) 012001 doi:10.1088/1742-6596/972/1/012001

Contamination and UV ageing of diffuser targets used in satellite inflight and ground reference test site calibrations

Anna Vaskuri[1,2], **Claire Greenwell**[2], **Isabel Hessey**[2], **Jordan Tompkins**[2], **and Emma Woolliams**[2]

[1]Metrology Research Institute, Aalto University, P.O. Box 15500, 00076 Aalto, Finland
[2]National Physical Laboratory, Teddington, TW11 0LW, United Kingdom

E-mail: anna.vaskuri@aalto.fi

Abstract. Diffuser reflectance targets are key components in in-orbit calibrations and for verifying ground reference test sites. In this work, Spectralon, Diffusil, and Heraeus diffusers were exposed to exhaust gases and ultraviolet (UV) radiation in the ambient air conditions and their degradations were monitored by measuring changes in spectral reflectances. Spectralon is a state-of-the-art diffuser made of polytetrafluoroethylene, and Diffusil and Heraeus diffusers are made of fused silica with gas bubbles inside. Based on the contamination tests, Spectralon degrades faster than fused silica diffusers. For the samples exposed to contamination for 20 minutes, the 250 nm – 400 nm total diffuse spectral reflectance of Spectralon degraded 3–5 times more when exposed to petrol-like emission and 16–23 times more when exposed to diesel-like emission, compared with Diffusil. When the reflectance changes of Spectralon were compared with those of Heraeus, Spectralon degraded 3–4 times more when exposed to petrol-like emission for 20 minutes and 5–7 times more when exposed to diesel-like emission for 7.5 minutes. When the samples contaminated were exposed to UV radiation in the ambient air, their reflectance gradually restored back to the original level. In conclusion, fused silica diffusers are more resistant to hydrocarbon contaminants present in ground reference test sites, and thus more stable under UV radiation in the air.

1. Introduction

Diffuser reflectance targets are key components in calibrating satellite sensors to support climate research, agriculture and forest monitoring [1, 2]. Inflight calibration systems use onboard diffusers as references, either as references for satellite Earth-reflectance products, or, combined with the known solar spectral irradiance, as references for satellite radiance products. Vicarious methods are also commonly used for satellite inflight calibration and validation. These use top-of-the-atmosphere reflectance models for reference test sites, e.g. RadCalNet [3], based on ground reflectance and atmospheric transmittance measurements and an atmospheric retrieval algorithm. Diffuser reflectance targets are also used in the characterisation and monitoring of such ground-based reference test sites. As reflectance targets are so critical for both onboard and vicarious satellite calibration it is essential to understand how these degrade in space and field environments. Degradation in both spectral reflectance and any changes in the bidirectional reflectance distribution function (BRDF) of the samples must be monitored. Unfortunately, space and field measurements involve contaminants, which can deteriorate the stability of the Spectralon diffusers currently used [1,4].

In this work Spectralon diffusers made of polytetrafluoroethylene (PTFE) are compared with alternative diffusers Diffusil and Heraeus, made of ultra-pure synthetic fused silica glass with gas bubbles. Especially Diffusil diffusers have been measured to have good BRDF properties [5]. Earlier studies have shown that pure Spectralon is photochemically stable and will not react with ultraviolet (UV) radiation [4]. However, if Spectralon contains residues e.g. of hydrocarbons, it becomes unstable under UV radiation [6]. In satellites, the diffusers are exposed to hydrocarbons due to the widely-used Pennzane lubricant [1] and possibly due to rocket fuel. In ground reference test sites, car emissions are a possible source of hydrocarbons [7].

The main objective of this paper is to find the best pre-ageing method for Spectralon targets, and to study alternative diffuser materials with more stable characteristics that could possibly replace Spectralon in the future. The second section introduces the setups for contaminating and UV ageing of the diffuser targets. The ageing results are discussed in Section 3. Finally, conclusions are drawn in Section 4.

2. Contaminating the samples

The setup used for contaminating the diffuser samples is shown in Fig. 1. The setup uses a propane flame produced using a mini-CAST soot source (JING mini-CAST Soot Generator Model 6203 Type C) as a particle emission source, which produces particles with composition and size similar to incomplete combustion of the fuel in car engines. The particle sizes (diameters) of the exhaust gases produced can be varied between 10 nm – 200 nm by mixing propane, nitrogen gas, oxidation air, and dilution air. The furnace at 400 °C can be used to remove volatile particles from the exhaust gas. Volatile particles have low boiling points and as the gas cools down, they condense on the surface of larger particles. The dilution air increases the velocity of the gas flow to encourage the particles to stick to the surface of the diffuser samples. Another purpose of the dilution air is to prevent volatile particles from recoalescing when the furnace is in use. A dilution loop in the setup is used for safety reasons as it allows particles to be filtered from gas flow when changing the samples.

Car emissions have wide particle size distributions; for diesel exhaust gases the peak of the distribution is ~100 nm, whereas for petrol exhaust gases it is ~20 nm [8]. Figure 2 shows the replicated distributions with the settings listed in Table 1. The particle sizes are determined by selection with a differential mobility analyser, which is calibrated with polystyrene latex spheres. The uncertainty of these sizes ($k = 2$) is about 5 %. Such particle emission distributions were used for contaminating the samples as detailed in Table 2. The particle size distributions in Fig. 2 are plotted at a logarithmic horizontal scale as the airborne particle emissions have been observed to follow near log-normal distributions [8]. From theoretical perspective, the log-normal particle size distribution can be derived from a random residence time approach [9].

IOP Conf. Series: Journal of Physics: Conf. Series **972** (2018) 012001 doi:10.1088/1742-6596/972/1/012001

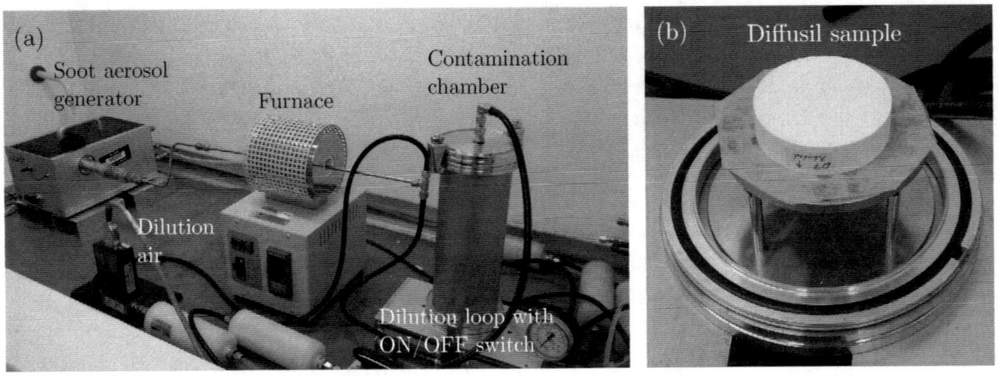

Figure 1. Contamination setup (a) and Diffusil sample on the holder of the chamber (b).

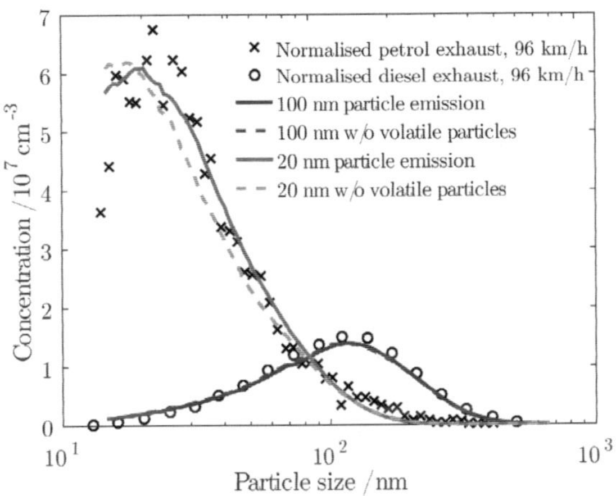

Figure 2. Particle size distributions with the concentrations of undiluted exhaust produced by the contamination setup to replicate petrol (crosses) and diesel (circles) exhaust measured in Ref. [8]. The petrol and diesel distributions [8] are normalised to the peaks of our particle size distributions.

Table 1. Settings for the gas inputs of the soot aerosol generator to produce petrol-like and diesel-like particle size distributions.

Type of emission	Particle size at the peak of the distribution nm	Propane ml/min	Nitrogen gas ml/min	Oxidation air l/min	Dilution air l/min
Petrol-like emission	20	20	10	0.6	2
Diesel-like emission	100	16	0	1	2

As listed in Table 2, we studied 10 Spectralon, 10 Diffusil, and 6 Heraeus samples. From each data sets, samples were exposed either to petrol-like or diesel-like particle emission. Diffusil samples have polished and rough sides and in this paper, we report the results for the rough sides. One Spectralon and one Diffusil sample were kept as uncontaminated references. Two Spectralon and two Diffusil samples were also exposed to the particle emissions with volatile particles evaporated from the exhaust gas by heating the gas with the furnace at 400 °C.

Table 2. Contamination times for Spectralon, Diffusil, and Heraeus sample sets. Petrol-like contamination refers to the 20 nm particle size distribution and diesel-like contamination refers to the 100 nm particle size distribution.

Diffuser type	Total number of samples	Sample ID	Type of particle emission	Exposure time / min
Spectralon USRS-99-020 by Labsphere	10	S1	Uncontaminated	–
		S2	Diesel-like	2.5
		S3	Diesel-like	5.0
		S4	Diesel-like	7.5
		S5	Diesel-like	20.0
		S6	Diesel-like, volatile particles removed	20.0
		S7	Petrol-like	5.0
		S8	Petrol-like	10.0
		S9	Petrol-like	20.0
		S10	Petrol-like, volatile particles removed	20.0
Diffusil R by Opsira	10	D1	Uncontaminated	–
		D2	Diesel-like	2.5
		D3	Diesel-like	5.0
		D4	Diesel-like	7.5
		D5	Diesel-like	20.0
		D6	Diesel-like, volatile particles removed	20.0
		D7	Petrol-like	5.0
		D8	Petrol-like	10.0
		D9	Petrol-like	20.0
		D10	Petrol-like, volatile particles removed	20.0
OM100 by Heraeus	6	H1	Diesel-like	2.5
		H2	Diesel-like	5.0
		H3	Diesel-like	7.5
		H4	Petrol-like	5.0
		H5	Petrol-like	10.0
		H6	Petrol-like	20.0

Degradation of the contaminated samples was monitored by measuring their 0°:45° spectral radiance factor and total diffuse spectral reflectance at 250 nm – 850 nm with a Perkin Elmer Lambda 900 spectrophotometer. To study how contaminated diffuser surfaces react to UV radiation, we radiated the samples gradually under the broadband spectrum of a xenon (Xe) lamp with the setup illustrated in Fig. 3 and monitored their degradation.

IOP Conf. Series: Journal of Physics: Conf. Series **972** (2018) 012001 doi:10.1088/1742-6596/972/1/012001

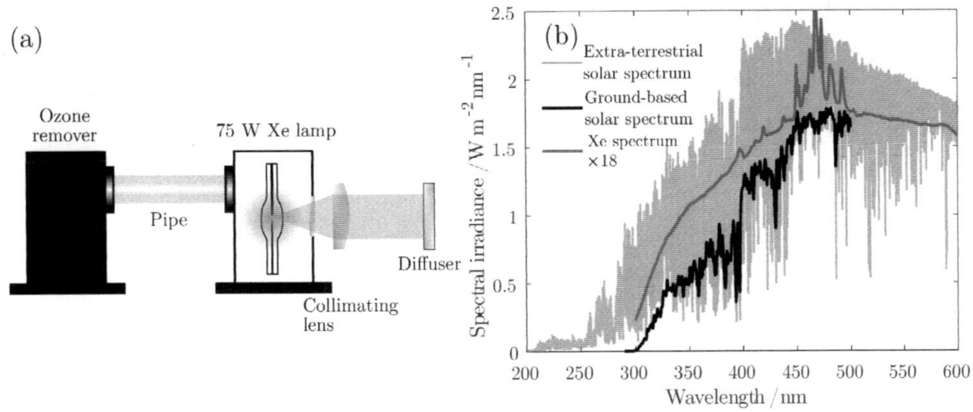

Figure 3. The setup for UV radiating the samples (a). The corresponding spectral irradiance on the sample plane was roughly 18 times lower in comparison with the ground-based solar spectrum measured at Izaña measurement campaign [10] (b). The extra-terrestrial solar spectrum has been determined by Gröbner *et al.* [11].

3. Results

3.1. Degradation due to contaminants

Spectralon is porous to hydrocarbon particles [1, 2, 4], which can be seen in Figs. 4(a) and 4(b) as a drastic drop in the total diffuse spectral reflectance at UV and visible (VIS) regions. The total diffuse spectral reflectance of an uncontaminated Spectralon reference, plotted as a black solid line, did not change. Contaminated Spectralon samples changed both in the UV and VIS. Diesel-like particle emission in Fig. 4(b) was more damaging to Spectralon than petrol-like particle emission in Fig. 4(a). In both cases, exhaust gas including volatile particles caused more degradation than that without volatile particles.

The total diffuse spectral reflectance of an uncontaminated Diffusil reference degraded 1 % at the UV region after 9 months, seen as a black solid line in Fig. 4(c). Fused silica diffusers change in the UV region due to hydrogen enrichment by defects, e.g. by non-bridging oxygen-hole and E'-centres [12]. The reflectances of Diffusil and Heraeus samples exposed to petrol-like particle emission also changed as seen in Figs. 4(c) and 4(e). Petrol-like particle emission was more damaging to these fused silica diffusers than diesel-like particle emission shown in Figs. 4(d) and 4(f). The reflectances of the Diffusil samples exposed to diesel-like particle emission did not change as systematically as those of the other samples. One possibility is that the ~100 nm particles did not stick to the surface and fell off when the samples were moved during the experiments. For fused silica diffusers, petrol-like emission was more damaging and in all these tests they were more resistant to contamination compared with Spectralon. The degradation of 0°:45° spectral reflectance was similar to the degradation of diffuse spectral reflectance for the diffuser samples studied.

IOP Conf. Series: Journal of Physics: Conf. Series **972** (2018) 012001 doi:10.1088/1742-6596/972/1/012001

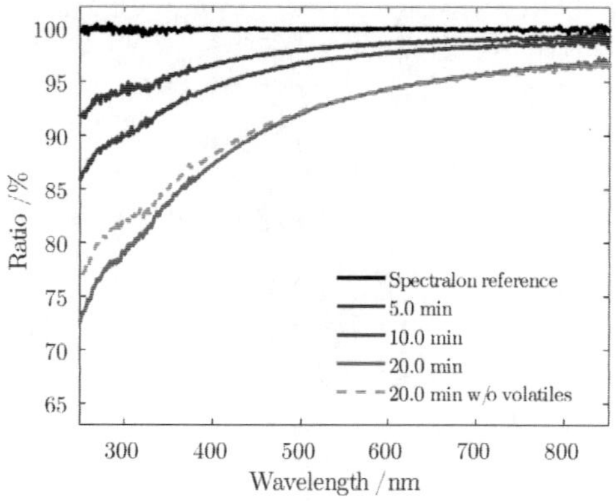

(a) ***Spectralon*** exposed to petrol-like emission

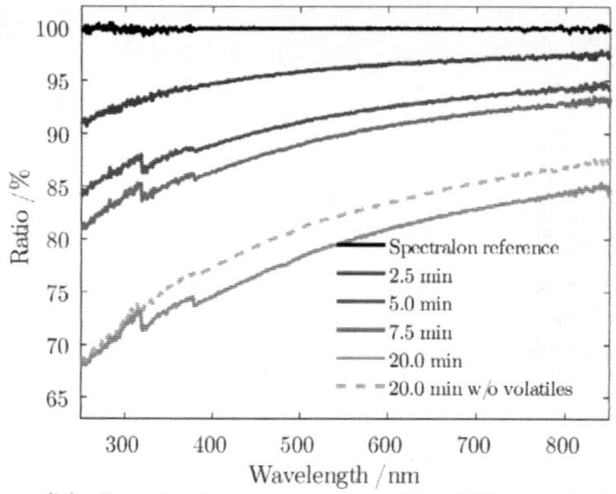

(b) ***Spectralon*** exposed to diesel-like emission

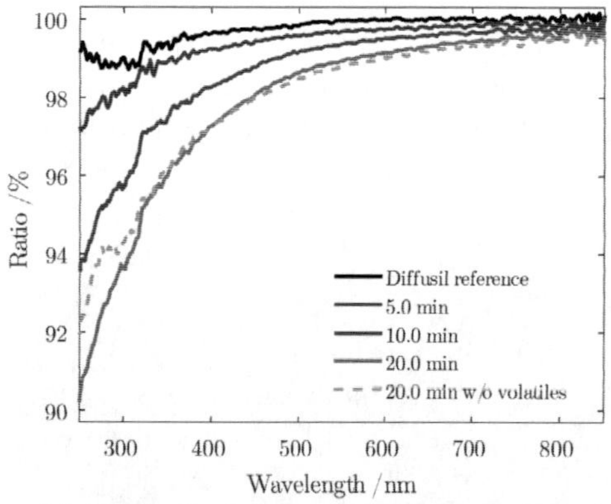

(c) ***Diffusil*** exposed to petrol-like emission

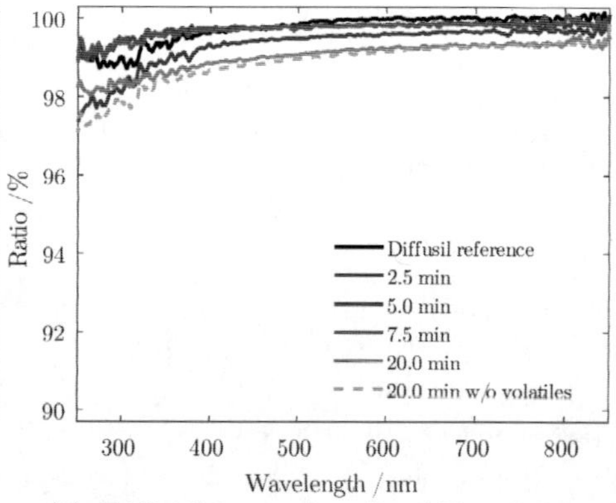

(d) ***Diffusil*** exposed to diesel-like emission

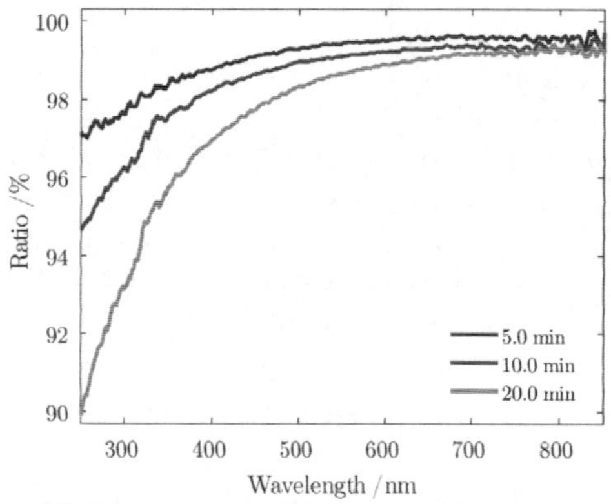

(e) ***Heraeus*** exposed to petrol-like emission

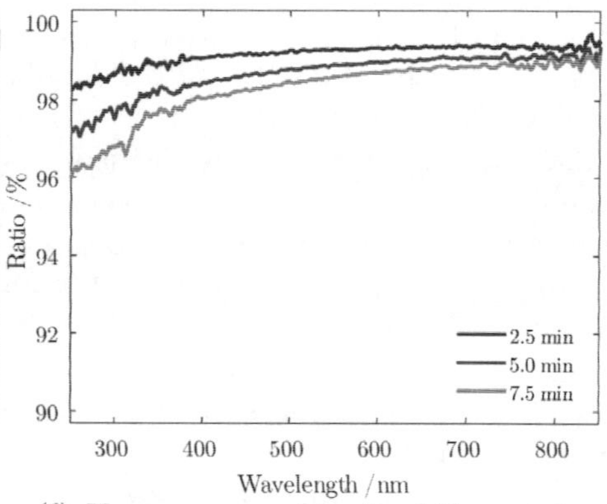

(f) ***Heraeus*** exposed to diesel-like emission

Figure 4. Ratios of the diffuse reflectances measured after and before the contamination for Spectralon (a)–(b), Diffusil (c)–(d), and Heraeus (e)–(f) samples. The degradation of the samples exposed to petrol-like particle emission are shown in figures on left and the degradation of the samples exposed to diesel-like particle emission are shown in figures on right. At this stage, the samples were not exposed to UV radiation.

3.2. Effect of combined contamination and UV radiation

We exposed the samples contaminated to the Xe lamp radiation with the setup illustrated in Fig. 3 and we noticed that UV radiation gradually cleaned the contaminated surfaces. Examples of such bleaching effect are presented in Fig. 5 for two Spectralon diffusers. The results for other diffuser types were systematically similar and the bleaching was relatively more effective for heavily contaminated diffuser surfaces.

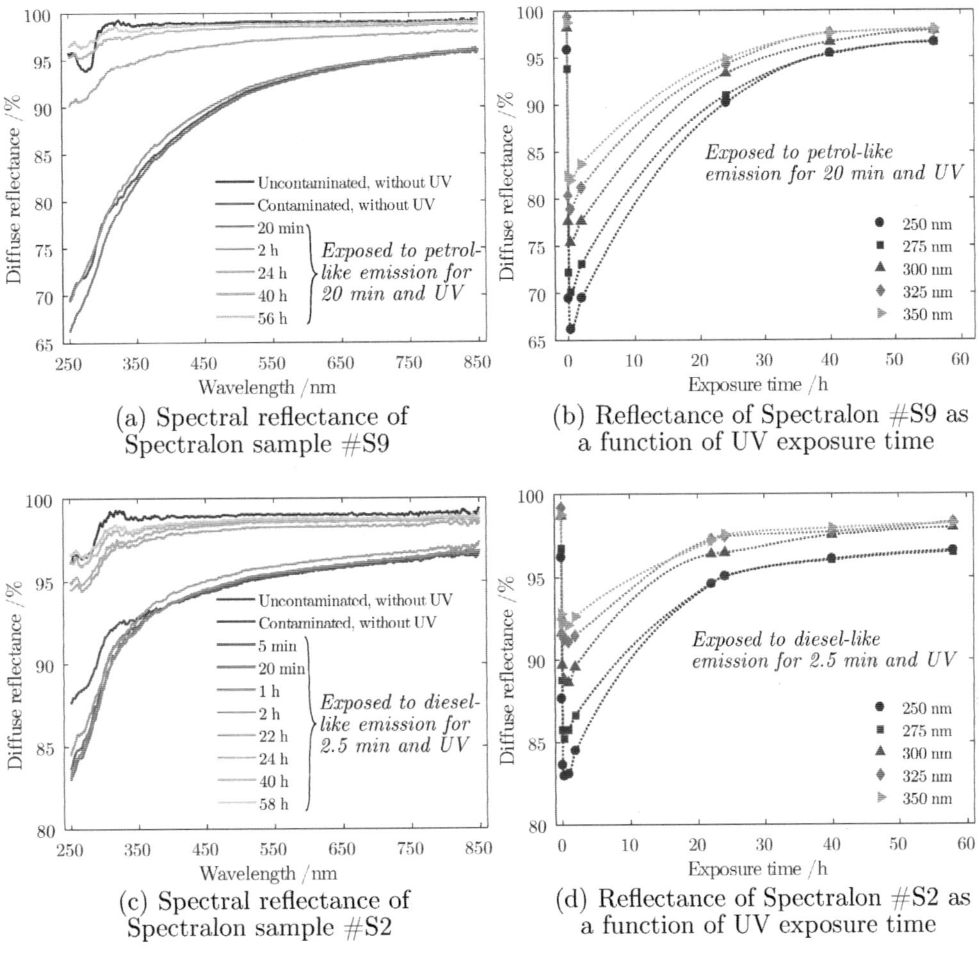

Figure 5. Examples of UV bleaching seen as the changing spectral reflectance of Spectralon samples; the sample #S9 exposed to petrol-like emission for 20 min and UV (a)–(b) and the sample #S2 exposed to diesel-like emission for 2.5 min and UV (c)–(d).

Based on our tests, mainly concerning ground reference test sites, the primary car emissions reduce reflectance. However, UV radiation in the presence of air bleaches the contaminated surfaces of Spectralon, Diffusil, and Heraeus diffusers. In the case of Spectralon, such bleaching in the air has also been noted in Ref. [13]. Bleaching takes place as UV radiation can break the relatively weak bonds of hydrocarbon molecules on the diffuser surface and the unstable hydrocarbons reach with the ambient air.

Based on literature in the case of in-orbit conditions [14], Spectralon is absorptive to vacuum UV radiation [15]. Such photon energies can break bonds between fluorine and carbon leading to the surface erosion of Spectralon [2, 16] and thus the surface loses fluorine [14, 17]. Broken carbon chains react with hydrocarbons outgassed, e.g. from lubricants [1], thus forming larger hydrocarbons. However, in the low-level orbit, atomic oxygen would bleach the surface [2, 14]. The atomic oxygen is formed when solar UVB and UVC radiation breaks the atmospheric oxygen molecules [18] according to Eq. (1). The atomic oxygen can further react with oxygen molecules and form ozone as shown in Eq. (2) that is also reactive. Top-of-the-atmospheric air is thin and sometimes the oxygen atom cannot find an oxygen molecule.

$$O_2 + \hbar\omega \longrightarrow 2O \quad (1)$$

$$O + O_2 \longrightarrow O_3 \quad (2)$$

In new Earth observation satellites, the problems related to the excess UV exposure of a diffuser target are reduced by having a diffuser wheel with a working diffuser and a reference diffuser [19]. The reference diffuser is only used for monitoring the degradation of the working diffuser and it is rarely exposed to UV radiation. However, the reference diffuser may still change e.g. due to contamination induced degradation.

4. Conclusions

Three diffuser technologies: Spectralon, Diffusil, and Heraeus, were exposed to hydrocarbon contaminants and UV radiation in the air. Hydrocarbon contamination degraded the reflectance of Spectralon more than they degraded fused silica diffusers. The 250 nm – 400 nm total diffuse spectral reflectance of Spectralon degraded 3–5 times more when exposed to petrol-like particle emission for 20 minutes and 16–23 times more when exposed to diesel-like particle emission for 20 minutes, compared with Diffusil. Comparing the degradation of Spectralon against that of Heraeus showed almost similar results; Spectralon degraded 3–4 times more when exposed to petrol-like particle emission for 20 minutes and 5–7 times more when exposed to diesel-like particle emission for 7.5 minutes.

When the samples contaminated were exposed to UV radiation in the air, their reflectance gradually restored back to the original level. This means that in the air, hydrocarbon degradation competes with UV bleaching and we could not develop an optimal pre-ageing method for Spectralon panels used in the ground reference test sites. The results presented in this paper are only indicative as we did not study the effect of some other contaminants, such as sand grains and secondary car emissions.

Based on the earlier studies of Spectralon in vacuum conditions, both UV radiation and contaminants caused reflectance degradation. Based on the findings presented in this paper, fused silica diffusers are more resistant to hydrocarbon contaminants present in ground reference test sites compared with Spectralon, and thus more stable under UV radiation in the air. However, further tests are needed to see how fused silica diffusers perform in vacuum conditions when exposed to hydrocarbon contaminants and UV radiation.

Acknowledgements

This work has been supported by the European Metrology Research Programme (EMRP) within the joint research project ENV53 "European metrology for Earth observation and climate" (MetEOC2). The EMRP is jointly funded by the EMRP participating countries within EURAMET and the European Union. Authors thank Paul Miller for helping with the UV irradiance measurements and Sara Pastor for helping with the reflectance measurements of the diffuser samples. Petri Kärhä is acknowledged for helping A.V. with the funding applications. A.V. thanks Jere Liukkonen for proof-reading this paper.

IOP Conf. Series: Journal of Physics: Conf. Series **972** (2018) 012001 doi:10.1088/1742-6596/972/1/012001

References

[1] Georgiev G T, Butler J J, Thome K J, Ramos-Izquierdo L A, Ding L, Graziani J, and Meadows G A, 2014 Initial studies of the directional reflectance changes in pressed and sintered PTFE diffusers following exposure to contamination and ionizing radiation *Metrologia* **51** S319–S328.

[2] Grossman E and Gouzman I, 2003 Space environment effects on polymers in low earth orbit *Nuclear Instruments and Methods in Physics Research B* **208** 48–57.

[3] Website: www.radcalnet.org

[4] Stiegman A E, Bruegge C J, and Springsteen A W, 1993 Ultraviolet stability and contamination analysis of Spectralon diffuse reflectance material *Optical Engineering* **32** 799–804.

[5] Lemaillet P, Patrick H J, Germer T A, Hanssen L, Johnson B C, and Georgiev G T, 2016 Goniometric and hemispherical reflectance and transmittance measurements of fused silica diffusers *Proceedings of SPIE* **9961** 996109.

[6] Bruegge C J, Stiegman A E, Ralnen R A, and Springsteen A W, 1993 Use of Spectralon as a diffuse reflectance standard for in-flight calibration of earth-orbiting sensors *Optical Engineering* **32** 805–814.

[7] Shaw P-S, Li Z, Arp U, and Lykke K R, 2007 Ultraviolet characterization of integrating spheres *Applied Optics* **46** 5119–5128.

[8] Harris S J and Maricq M M, 2001 Signature size distributions for diesel and gasoline engine exhaust particulate matter *Aerosol Science* **32** 749–764.

[9] Kiss L B, Söderlund J, Niklasson G A, and Granqvist C G, 1999 The real origin of lognormal size distributions of nanoparticles in vapor growth processes *Nanostructured Materials* **12** 327–332.

[10] Vaskuri A, Kärhä P, Egli L, Gröbner J, and Ikonen E, (submitted) Monte Carlo method for determining uncertainty of total ozone derived from direct solar irradiance spectra: Application to Izaña results.

[11] Gröbner J, Kröger I, Egli L, Hülsen G, Riechelmann S, and Sperfeld P, 2017 The high resolution extra-terrestrial solar spectrum determined from ground-based solar irradiance measurements *Atmospheric Measurement Techniques* **10** 3375–3383.

[12] Nürnberg F, Kühn B, Langner A, Altwein M, Schötz G, Takke R, Thomas S, and Vydra J, 2015 Bulk damage and absorption in fused silica due to high-power laser applications *Proceedings of SPIE* **9632** 96321R.

[13] Dever J A, Rutledge S K, Hasegawa M M, and Reed C K, 1998 Evaluation of Low Earth Orbit Environmental Effects on International Space Station Thermal Control Materials *NASA / TM-1998-208500*, 11 pages.

[14] Zhao X, Shen Z, Xing Y, and Ma S, 2004 Experimental Study of Vacuum Ultraviolet Radiation Effects and Its Synergistic Effects with Atomic Oxygen on a Spacecraft Material Polytetrafluoroethylene *Chinese Journal of Aeronautics* **17** 181–186.

[15] Townsend J A, Hansen P A, McClendon M W, de Groh K K, and Banks B A, 1999 Ground-based testing of replacement thermal control materials for the Hubble Space Telescope *High Performance Polymers* **11** 63–79.

[16] Ishikawa K, Sumi N, Kono A, Horibe H, Takeda K, Kondo H, Sekine M, and Hori M, 2012 Synergistic Formation of Radicals by Irradiation with both Vacuum Ultraviolet and Atomic Hydrogen: a Real-time in situ Electron Spin Resonance Study, https://arxiv.org/pdf/1206.6920.

[17] Ferry L, Vigier G, Alexander-Katz R, and Garapon C, 1998 Interaction between UV Radiation and Filled Polytetrafluoroethylene (PTFE). I. Degradation Processes *Journal of Polymer Science B* **36** 2057–2067.

[18] Fabry C and Buisson H, 1921 A study of the ultra-violet end of the solar spectrum *Astrophysical Journal* **54** 297–322.

[19] Pérez Albiñana A, Erdmann M, Wright N, Martin D, Melf M, Bartsch P, and Seefelder W, 2017 Sentinel-5: the new generation European operational atmospheric chemistry mission in polar orbit *Proceedings of SPIE* **10403** 104030P.

Setup for studying speckle noise of spectroradiometer diffusers in Earth observation applications

Anna Vaskuri[1,2], Claire Greenwell[2], and Emma Woolliams[2]

[1]Metrology Research Institute, Aalto University, P.O. Box 15500, 00076 Aalto, Finland
[2]National Physical Laboratory, Teddington, TW11 0LW, United Kingdom

E-mail: anna.vaskuri@aalto.fi

Abstract. Diffusers in in-orbit spectroradiometers cause speckle under partially coherent solar radiation. A speckle pattern entering a spectroradiometer through a small slit creates systematic spectral deviations in measured spectra. We have developed a setup to characterise the spatial speckle of diffusers and the related spectral features. The decorrelation angles measured at 532 nm for Spectralon, Diffusil, and Heraeus diffusers were 0.021°, 0.014°, and 0.005° respectively. This information can be used for compensating speckle-related spectral features from the radiometric satellite measurements by averaging over multiple decorrelated spectra.

1. Introduction

Diffuser speckle is the result of constructive and destructive interference arising from the surface roughness [1,2]. Speckle noise can be observed only under fully or partially coherent illumination produced by a small or point source, e.g. by a laser or the Sun observed from a long distance [2]. Such effect is called spatial speckle. Speckle can also be temporal, but this usually occurs only in the beam of narrow-band gas lasers. An example of speckle measured on a Spectralon diffuser is presented in Fig. 1.

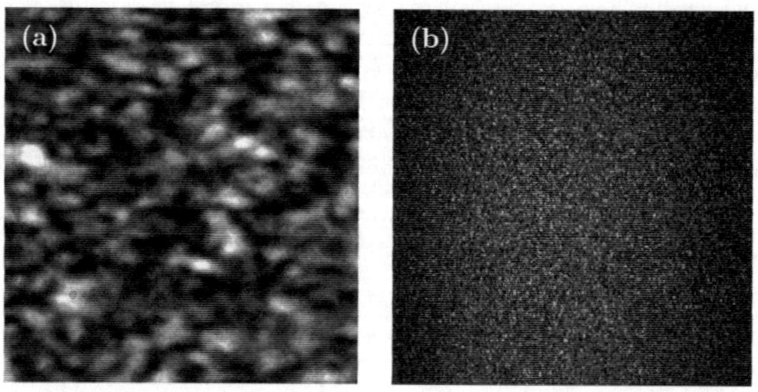

Figure 1. Spatial speckle noise measured for Spectralon. In (a), the speckle image of 600 × 600 px measured in free-space propagation mode with a source spot size of 5 mm. In (b), the image of 250 × 250 px measured in imaging mode with $f/22$. The edges in the image (b) appear darker due to the Gaussian spatial profile of the laser beam.

In high-resolution satellite spectroradiometers, the diameter of the entrance slit is tens of micrometres [1, 2]. Thus, the diffraction pattern arising from diffuser roughness creates systematic spectral deviations, i.e. spectral features, in measured spectra. The sign of such deviations depends on the number of constructive and destructive speckles that enter through the slit. Speckle is a small component in the uncertainty budget of in-orbit calibration and it is more enhanced towards the near infrared (NIR) region [3]. The main problem with the speckle is that it leaves slowly-varying systematic deviations in the measured spectra.

2. Measurement setup

The developed speckle setup is illustrated in Fig. 2, and is based on a similar earlier setup developed by the Netherlands Organisation for Applied Scientific Research (TNO) [1, 4]. Our setup is automated and controlled with custom-made LabVIEW software including measurement sequences to determine spatial speckle and the spectral features caused by speckle. A diffuser is mounted in the centre of a motorised rotary stage. To mimic the measurement conditions of in-orbit calibration, the diffuser sample and a camera (or a spectroradiometer) are stationary with respect to each other while the angle of the incident radiation changes. Speckle reflected by the diffuser is measured with a 2/3" silicon (Si) CCD camera (Basler, 2448×2050 px) with an imaging lens by Sigma, by illuminating the diffuser with a 532 nm diode laser or a wavelength-tuneable supercontinuum laser [5]. The camera system was not corrected for the flat-field. We used the area of 200×200 px in the middle of the array for calculating the angular speckle correlation functions, and thus the vignetting of the camera system was negligible. The available spectral range is from 400 nm to 1000 nm, limited by the responsivity of the Si camera and the wavelength-tuneable laser. The polarisation of the laser beam can be selected by rotating a polariser. In addition, the setup has the option to attenuate the beam by a neutral density filter to avoid the saturation of the camera. A motorised shutter is used for measuring and subtracting the dark signal of the camera and straylight from the images. The speckle pattern can be measured at various angles. The laser beam is expanded with an achromatic beam expander so that the beam with the diameter ($1/e^2$) of 1 cm hits to the diffuser surface. Beam profile monitoring in Fig. 2(b) also confirmed the absence of temporal laser speckle from the laser sources used.

Spectral features with respect to different incident angles of radiation are measured by illuminating the diffuser with the monochromatic laser beam, taking images, summing pixel intensities inside the selected area of the image and then calculating percentage difference between summed intensities. These steps are repeated at each wavelength. The alternative, more straightforward method to measure systematic spectral variations arising from speckle would be to use the white light of the supercontinuum laser without the diffraction filter system and measure the spectral changes with a spectroradiometer, but we did not do that due to safety reasons. The partially coherent white light could also be obtained by a collimated xenon (Xe) lamp radiation.

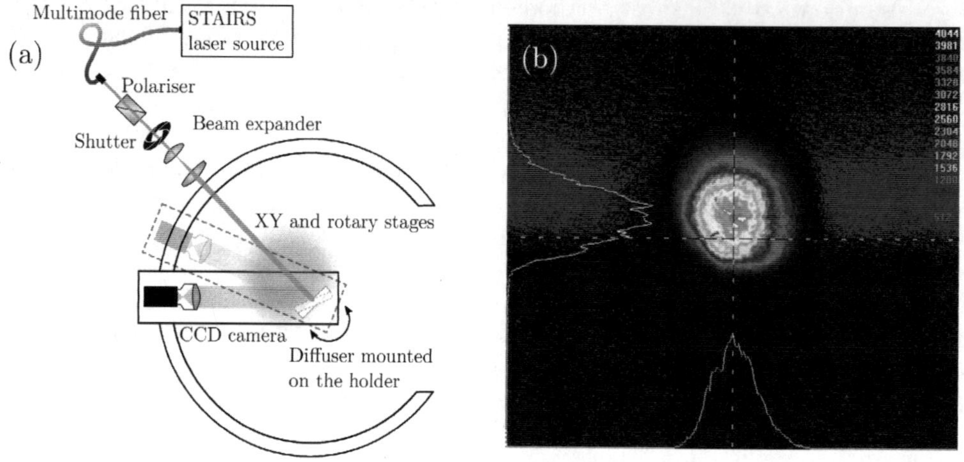

Figure 2. Illustration of the speckle measurement setup with STAIRS refering to the wavelength-tuneable supercontinuum laser (a). The STAIRS with multimode fiber or 532 nm diode laser could be used as a light source. The spatial beam profile of the 532 nm diode laser is shown in (b).

3. Results and discussion

Speckle-related spectral features in satellite spectroradiometer measurements can be attenuated effectively by averaging over the spectra measured at slightly different angles [1,2]. To determine the required decorrelation angle, correlation coefficients γ between each image pair measured at angles θ_1 and θ_2 are calculated as [4]

$$\gamma(\theta_1, \theta_2) = \frac{\sum_{i=1}^{M}\sum_{j=1}^{N}\left(I_{\theta_1,x_{ij}} - \bar{I}_{\theta_1}\right)\left(I_{\theta_2,x_{ij}} - \bar{I}_{\theta_2}\right)}{\sqrt{\sum_{i=1}^{M}\sum_{j=1}^{N}\left(I_{\theta_1,x_{ij}} - \bar{I}_{\theta_1}\right)^2 \sum_{i=1}^{M}\sum_{j=1}^{N}\left(I_{\theta_2,x_{ij}} - \bar{I}_{\theta_2}\right)^2}}. \tag{1}$$

Two images are considered decorrelated when the correlation coefficient drops below 0.5. Figure 3(a) shows the decorrelation angles of Spectralon, Diffusil, and Heraeus diffusers measured under 532 nm coherent radiation (0.021°, 0.014° and 0.005° respectively). Diffusil and Heraeus samples are made of fused silica glass with gas bubbles inside and were measured on both sides. The decorrelation angles are wavelength dependent and they increase towards longer wavelengths.

Decorrelation angles $\delta\theta$ give information about the structure of the material, e.g. their bubble or grain size σ, since a simplified analytical expression for the correlation function is [6]

$$\gamma(\sigma, \delta\theta) = \exp\left(-\left[(2\pi n/\lambda)\sigma \sin(\theta_i)\,\delta\theta\right]^2\right), \tag{2}$$

where n is the refractive index of medium, λ is the laser wavelength and θ_i is the angle of incident radiation. The roughness levels estimated from the decorrelation angles using Eq. (2) are ~3.3 µm, ~5.2 µm, and ~15 µm, for Spectralon, Diffusil and Heraeus diffusers, respectively. The corresponding literature values from Refs. [7–9] state the roughness of ~2 µm for Spectralon, ~3 µm for Diffusil, and <20 µm for Heraeus. The model of Eq. (2) was originally developed for surface diffusers, and thus the Gaussian models (dotted lines) do not fit well to the correlation functions (symbols) for our volume diffusers in Fig. 3(a). To model the speckle correlation for volume diffusers more accurately, a complex correlation model [10–13] has to be used. This model is fitted to the experimental data of Fig. 3(a) as solid lines. In the model, the roughness σ is proportional to the mean photon path length l in the optical medium. Example results for spectral features, arising from the speckle are measured for Spectralon in Fig. 3(b).

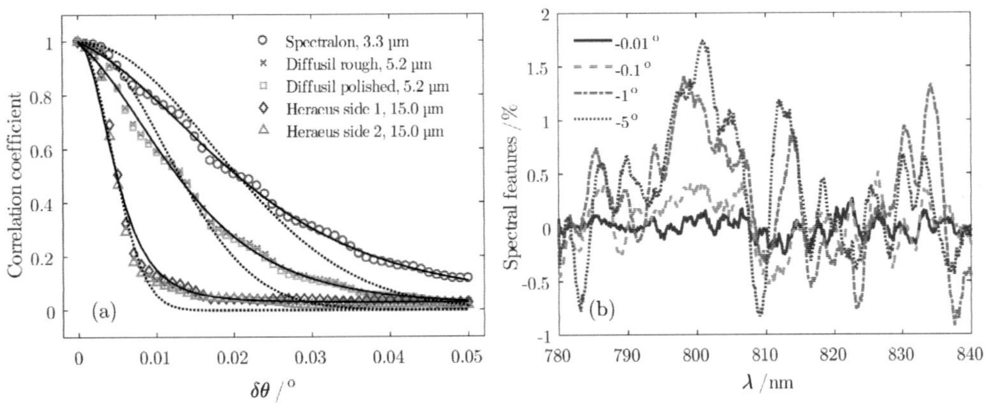

Figure 3. Correlation coefficients as a function of angular change for three diffusers (a) and spectral features measured for Spectralon (b). In (a), correlation functions as symbols measured at 532 nm coherent radiation, a simple model of Eq. (2) as dotted lines and a model for volume diffusers as solid lines [11, 13].

When considering the speckle effect only, the diffuser with the smallest speckle decorrelation angle would be the best. However, many other phenomena affect the diffuser performance more than speckle, such as the bidirectional reflectance distribution function (BRDF) of the diffuser and how well it can be modelled, and the degradation of the diffuser over time. Thus, selecting the diffuser target is always a compromise between speckle properties, BRDF and ageing of the diffuser material.

4. Conclusions

In this work, we have developed a setup to characterise the spatial speckle of diffusers and the related spectral features. The setup can be operated with a wavelength-tuneable supercontinuum laser or e.g. with a diode laser within the spectral range of 400 nm – 1000 nm. The decorrelation angles measured at 532 nm for Spectralon, Diffusil, and Heraeus diffusers were 0.021°, 0.014°, and 0.005° respectively. The speckle-related spectral features can be compensated from the radiometric satellite measurements by averaging over multiple decorrelated spectra.

Acknowledgements

This work has been supported by the European Metrology Research Programme (EMRP) within the joint research project ENV53 European metrology for Earth observation and climate (MetEOC2). The EMRP is jointly funded by the EMRP participating countries within EURAMET and the European Union. A.V. is grateful for the travel grant by the Finnish Foundation for Technology Promotion. Lydia Zajiczek, Paul Miller, Paul Green, Barry Scott and Florian Graber are acknowledged for their technical feedback and for sharing their expert knowledge in optics. Authors thank Petri Kärhä for helping with the funding applications.

References

[1] van Brug H and Bazalgette Courrèges Lacoste G, 2007 Spectral features, effects and cures *Proceedings of SPIE* **6677** 667703.

[2] van Brug H H, Vink R, Groote Schaarsberg J, Bazalgette Courrèges-Lacoste G, and Snijders B, 2004 Speckles and their effects in spectrometers due to on-board diffusers *Proceedings of SPIE* **5542** 334–341.

[3] Gulde S T, Kolm M G, Smith D J, Maurer R, Bazalgette Courrèges-Lacoste G, Sallusti M, and Bagnasco G, 2014 Sentinel 4: A geostationary imaging UVN spectrometer for air quality monitoring Status of design, performance and development *Proceedings of International Conference on Space Optics (ICSO)*.

[4] Gür B, van Brug H, Xu M, and Vela E, 2014 Diffuser properties and according performance in BSDF and spectral features in remote sensing applications *Proceedings of SPIE* **9245** 92450I.

[5] Levick A P, Greenwell C L, Ireland J, Woolliams E R, Goodman T M, Bialek A, and Fox N P, 2014 Spectral radiance source based on supercontinuum laser and wavelength tunable bandpass filter *Applied Optics* **53** 3508–3519.

[6] Léger D, Mathieu E, and Perrin J C, 1975 Optical Surface Roughness Determination Using Speckle Correlation Technique *Applied Optics* **14** 872–877.

[7] Parretta A and Addonizio M L, 2015 Optical and Structural Characterization of Diffuse Reflectance Standards *International Journal of Optics and Applications* **5** 33–49.

[8] Heraeus, *OM 100 Opaque Quartz Glass*, Datasheet 2007.

[9] Opsira, *Diffusil – The perfect Lambertian quartz diffusers*, Datasheet 2014.

[10] Okamoto T and Yanagita S, 2003 Intensity correlations of coherent light reflected from dense powders *Optics Communications* **227** 15–23.

[11] Akkermans E, Wolf P E, Maynard R, and Maret G, 1988 Theoretical study of the coherent backscattering of light by disordered media *Journal de Physique* **49** 77–98.

[12] Lagendijk A, Vreeker R, and De Vries P, 1989 Influence of internal reflection on diffusive transport in strongly scattering media *Physics Letters A* **136** 81–88.

[13] Okamoto T and Yanagita S, 2001 Correlation properties of light multiple scattered from powders *Proceedings of SPIE* **4416** 440–443.

Applying Metrological Techniques to Satellite Fundamental Climate Data Records

Emma R Woolliams[1], Jonathan PD Mittaz[1,2], Christopher J Merchant[2], Samuel E Hunt[1] and Peter M Harris[1]

[1]National Physical Laboratory, Teddington, UK,
[2]University of Reading, Reading, UK

Corresponding e-mail address: emma.woolliams@npl.co.uk

Abstract. Quantifying long-term environmental variability, including climatic trends, requires decadal-scale time series of observations. The reliability of such trend analysis depends on the long-term stability of the data record, and understanding the sources of uncertainty in historic, current and future sensors. We give a brief overview on how metrological techniques can be applied to historical satellite data sets. In particular we discuss the implications of error correlation at different spatial and temporal scales and the forms of such correlation and consider how uncertainty is propagated with partial correlation. We give a form of the Law of Propagation of Uncertainties that considers the propagation of uncertainties associated with common errors to give the covariance associated with Earth observations in different spectral channels.

1. Introduction

Satellite-derived Earth Observation (EO) data provide a legacy of information about environmental and climate changes that is of immense value to society. Obtaining information about long-term trends requires the comparison of satellite-derived observations taken decades apart, using different sensors and often the combination of data from sensors measuring in, e.g. different regions of the electromagnetic spectrum. To ensure that such comparisons are meaningful, it is essential to quantify the stability of satellite sensors and to determine the radiometric differences between different sensors and their associated uncertainties.

Some satellite series have been available for almost 40 years, the first versions of which were launched in the late 1970s. Each satellite had a limited lifetime, and therefore a series of such satellites were launched, often with identical sensors on board, or with newer instrument generations that were an evolution of earlier designs. The operational calibrations of such sensor series were not intended for climate applications – for example, they are not stable over time. However, they made measurements over a time of significant climate change and the records from their observations, though currently not fit for purpose, are of great interest for our understanding of the Earth's environment.

Metrological principles of traceability, uncertainty analysis and comparison can be used to quantify long-term stability of Fundamental Climate Data Records (FCDRs); i.e. multi-decadal observations of fundamental quantities (such as top-of-atmosphere Earth radiance) from multi-satellite series. However, techniques established for laboratory calibrations cannot translate directly into Earth observation applications. The process of launch and natural degradation in space, along with the lack of SI-traceable references in orbit means that traceability is lost after launch. With historical sensors, particularly those launched in the 1980s – 90s, there is often also insufficient accurate information available from the pre-flight calibrations to perform a complete uncertainty analysis. Furthermore, in orbit, repeat measurements are generally not available since the turbulent and chaotic atmosphere modifies the top of atmosphere radiance observed from space.

In this paper we discuss approaches defined within the FIDUCEO project [1] for applying metrological techniques to FCDRs in the visible, thermal infrared and microwave spectral regions, We also introduce some concepts for the next step: the application of these FCDRs to generate Climate Data Records (CDRs) of geophysical products, with as rigorous uncertainty analysis. This paper is a conference proceedings paper and the ideas will be further elaborated in a paper in preparation [2].

Considered also is how we can provide a better calibration of the instrument, based on what we know now, and how we can establish more rigorous uncertainty analysis for the observations from such sensors in order to develop FCDRs. The recalibration process, harmonisation, involves the determination of new calibration coefficients for the instrument, either to improve the information available from pre-flight calibration, or to estimate quantities that were not determined pre-flight. Harmonisation is achieved using match ups: where a particular satellite sensor observed the same Earth scene as another sensor at the same time and in similar spectral bands/viewing angles.

The development of an FCDR then involves three main steps: (1) the establishment of a measurement equation, which is used to convert the observed counts into the FCDR quantity (e.g. radiance); (2) the analysis of the uncertainty associated with each quantity in the measurement equation, and the error correlation structure for the effects that influence each quantity; and (3) the harmonisation of the series. In the following sections we discuss these steps.

2. The measurement equation

We start our analysis of an FCDR with the measurement equation for the sensor, which converts the measured detector counts (the Level 0 product) into radiance or reflectance (the Level 1 product). It may be appropriate to use a different measurement equation for historical reanalysis than that which was used for the production of operational data, allowing for additional effects to be taken into account. For example for the AVHRR (Advanced Very High Resolution Radiometer) instrument, onboard the NOAA (National Oceanic and Atmospheric Administration, USA) satellite series since 1978, the (simplified) measurement equation [3] takes the form

$$L_{\mathrm{E}} = a_0 + \frac{(\varepsilon + a_1) L_{\mathrm{ICT}} - a_2 \tilde{C}_{\mathrm{ICT}}^2}{\tilde{C}_{\mathrm{ICT}}} C_{\mathrm{E}} + a_2 C_{\mathrm{E}}^2 + 0 \tag{1}$$

where L_{E} is the Earth radiance observed by a particular instrument channel and for a particular image pixel; C_{E} is the measured Earth count for that pixel; L_{ICT} is the radiance of the internal calibration target (ICT) in that channel; $\tilde{C}_{\mathrm{ICT}}$ is the measured (averaged) signal from observing that ICT; ε is the emissivity; the different a_i are harmonised calibration parameters and 0 represents effects relating to the assumptions implicit in the form of the measurement equation. In general, the calibration parameters will be estimated through a combination of pre-flight calibration (here, emissivity ε), in-flight calibration (in this case the gain – the term in front of the linear counts term – which is determined from the ICT radiance and counts) and, for historical sensor series, through harmonisation (here, the a_i). Harmonisation is the process of determining improved calibration parameters for a series of satellite sensors by using information available from comparisons of those sensors with each other and with a reference through 'match-ups': occasions when the different sensors saw the same scene at the same time due to satellite overpasses. We explicitly include the + 0 term in the measurement equation to highlight the inbuilt assumptions. Here, for example, the equation includes a quadratic nonlinearity, but are assuming there are no higher nonlinearities; we are also making assumptions of sufficient monochromaticity as many terms are band-integrated quantities.

3. Uncertainty effects and correlation structures

The measurement equation, Eq.(1), describes the Level 1 product (here radiance) for a particular image pixel. Each term in the measurement equation is sensitive to a number of uncertainty effects.

Each of these effects has an associated uncertainty and will give rise to an unknown error, which can be considered a draw from a probability distribution whose standard deviation is given by the standard uncertainty. While the error itself cannot be known, we can say something about the error correlation between the measured Earth radiance values in different pixels of a satellite sensor image, or between different spectral bands on the satellite sensor. Independent random effects will cause errors which can never be corrected for in a single measured value, even in principle, because the effect is stochastic. In EO, a common example of a random effect is radiometric noise. In the laboratory, the uncertainty associated with random effects is usually determined by calculating the standard deviation of values from repeated measurements. For Earth observations, this method is only available for onboard calibration references and the noise for Earth scenes must be deduced from these calibration measurements and an understanding of the physics of the sensor.

Not all random effects are independent. Structured random effects arise in EO when effects are random, but have predictable patterned relationships across a set of measurements, perhaps because values of the effects are common for multiple pixels. An example is a cross-track sensor for which calibration measurements against a reference target is performed once per scanline. Any random effects associated with the derived calibration affect all Earth observations in that scan and therefore provide a common error to all those pixels.

Systematic effects are those that could in principle be corrected for, if sufficient knowledge about the effect were to be available, although even after any correction there is always some residual systematic error that remains that is unknown and quantified only by its associated uncertainties. These systematic effects cause errors that can be related across multiple measurements, perhaps for a whole mission. For example an error in the measurement of the instrument spectral response function will affect the analysis of all measurements by that sensor, though not in the form of a simple bias.

Because FCDRs are processed to produce CDRs (Level 2) by combining results from different spectral bands, and gridded-CDRs (Level 3) through the spatial and/or temporal combination of values in different pixels, it is important to determine the error-correlation structure between spectral bands and between different pixels. Note that the error correlation structure can be different in different dimensions. For example, a structure random effect caused by a calibration performed once per scanline, will be fully correlated between any two pixels within a scanline, but uncorrelated between any two pixels in different scanlines. Similarly an effect that depends on mirror angle may be fully correlated for pixels in the same relative position in all scanlines, but uncorrelated from one pixel to another within a scanline. For low Earth orbit satellites with across-track scanning (sensors that observe the Earth scanline by scanline as the satellite passes along a track), the most sensible dimensions over which to consider error correlations are within a scanline, from scanline to scanline and from orbit to orbit. Separately we must consider the correlation structure from channel to channel.

3.1. Error correlation forms

Error correlation occurs when there is a common error between two observations (in different spectral bands and/or in different spatial/temporal positions). In general, the error correlation form can be determined by considering the pattern for the common error. While there may be a large number of possible error correlation forms, we have defined a small subset which we consider sufficient for representing most practical cases for satellite sensors. Note that a particular effect may have a different correlation structure in different dimensions, as given above.

The "independent random" correlation represents the simplest case, when there is no error correlation for any pair of measured values (in the dimension of interest). A common example is noise on the Earth pixel count.

The "rectangular absolute" correlation is an error correlation form where there is a common error over a range of values defined in an absolute sense. A fully systematic effect would have a rectangular absolute correlation structure of infinite range. A finite range would represent a situation where a common error affects the measured values for a finite number of pixels. For example, if a single calibration value is obtained and used for all measurements within a particular time period, or for a certain number of scanlines, and this calibration value is then replaced with a refreshed value for a subsequent time period or set of scanlines, then there is a full correlation between values for pixels within that calibration period, and no correlation between values for pixels from one calibration period to the next.

The "triangular relative" correlation is an error correlation that arises from a running average with equal weights. If, for example, a calibration obtained once per scanline is averaged in a simple running average over a certain number of scanlines before and after the considered scanline, then the error correlation structure from one scanline to another would depend on the number of common calibrations. This provides a triangular correlation structure from scanline to scanline with the full base equal to twice the number of scanlines averaged. (Note that in this example, the correlation structure from pixel to pixel within the scanline is a rectangular absolute, across the whole scanline).

The "bell-shaped relative" correlation is an error correlation that occurs from a weighted average with a larger weight applied to the central value. The correlation structure can be considered as the convolution of two functions represented by the weights assigned to the rolling average. For a triangular weighting function this takes the form of a cubic spline over the overlap range (full base equal to twice the number of scanlines involved in the average). This cubic spline is sufficiently close to a Gaussian distribution in shape that we have represented a "bell-shaped relative" as a truncated Gaussian shape. For weights other than a triangle, the bell-shape is also likely to be sufficiently represented by a truncated Gaussian shape. This error correlation structure can also represent effects where we have very little information but the recognition that points closer in space/time are more correlated than those further away.

Within the FIDUCEO project we have developed a framework for formally coding this information, intend to be stored computationally in NetCDF files, this will be discussed further in future work.

3.2. Propagation of uncertainties

3.2.1. Uncertainties for a single pixel of an FCDR

In most cases we can consider each effect as independent of other effects. That means that the only covariance we need to consider for determining the Earth radiance of a single pixel (the FCDR quantity), is the covariance associated with the different harmonisation coefficients (see Section 2.2). Therefore we can write the Law of Propagation of Uncertainty [3] as

$$u^2\left(L_{\mathrm{E}}\right)=\sum_{i=1}^{n}\left(\frac{\partial L_{\mathrm{E}}}{\partial x_i}\right)^2 u^2\left(x_i\right)+\sum_{j=1}^{m}\left(\frac{\partial L_{\mathrm{E}}}{\partial a_j}\right)^2 u^2\left(a_j\right)+2\sum_{j=1}^{m-1}\sum_{k=j}^{m}\left(\frac{\partial L_{\mathrm{E}}}{\partial a_j}\right)\left(\frac{\partial L_{\mathrm{E}}}{\partial a_k}\right)u\left(a_j,a_k\right) \qquad (2)$$

where the first term relates to the effects in the effects table, added in quadrature (the sensitivity coefficients and uncertainties are given in the effects tables); the second term relates to the uncertainty associated with each harmonisation coefficient; and the third term relates to the covariance of the harmonisation coefficients. Note that currently this level of uncertainty propagation is not undertaken at the radiance level for most if not all satellite products.

IOP Conf. Series: Journal of Physics: Conf. Series **972** (2018) 012003 doi:10.1088/1742-6596/972/1/012003

3.2.2. Error covariance for different channels in a single pixel

Often for the establishment of a CDR (climate data record of a geophysical parameter) from an FCDR (Earth radiance values used for CDR generation) the measured values in different spectral channels are combined. Therefore it is necessary to determine the error covariance between the Earth radiance values obtained for a single pixel in multiple spectral channels, and to provide a correlation matrix, $\boldsymbol{R}$, representing, for that effect, what is the correlation between channels. This matrix will often, but not always, be either the identity matrix or a matrix having unity in every element depending on whether an effect is common from channel to channel or has an independent error from channel to channel. Note that this is different from the error correlation structures describing pixel-to-pixel correlation, see following section.

For each effect in turn, we can consider the error covariance matrix for Earth radiance between channels due that that effect, to be the matrix multiplication of simple matrices, each square with a dimension equal to the number of channels. The matrix $\boldsymbol{C}_k$ has the sensitivity coefficients for each channel's Earth radiance to the quantity of the effect down the diagonal (and zero values elsewhere). The matrix $\boldsymbol{U}_k$ has the standard uncertainties associated with the effect (in units of the quantity) down the diagonal, and zeroes elsewhere. Note that for effects that are correlated between channels, this will be the same value down all elements of the diagonal, for effects that are not correlated between channels, it may have a different value down each element of the diagonal. The matrix $\boldsymbol{R}_k$ is the channel-to-channel correlation matrix for this effect. The covariance matrix for Earth radiance from channel-to-channel due to a single effect is given by a multiplication of these matrices.

As an example, consider the situation with three spectral channels that are all calibrated with the same onboard blackbody calibrator (ICT). The unknown error in the temperature of that blackbody, a draw from the probability distribution described by the uncertainty associated with the blackbody temperature, will be a common error for all three channels. Therefore the matrix $\boldsymbol{R}_i$ is a full matrix of ones. The $\boldsymbol{V}_i$ has the standard uncertainties associated with temperature down the diagonal (say 10 mK) and the $\boldsymbol{C}_i$ matrix has the sensitivity coefficient of each channel's Earth radiance ($L_{\mathrm{E},c}$) to the uncertainty associated with temperature, which will likely be calculated in a chain rule as the partial derivative of Earth radiance with respect to the calibration target radiance and the partial derivative of the ICT radiance with respect to temperature: $\frac{\partial L_{\mathrm{E},c}}{\partial T} = \frac{\partial L_{\mathrm{E},c}}{\partial L_{\mathrm{ICT},c}} \frac{\partial L_{\mathrm{ICT},c}}{\partial T}$. Thus

$$\boldsymbol{C}_T\boldsymbol{U}_T\boldsymbol{R}_T\boldsymbol{U}_T^{\mathrm{T}}\boldsymbol{C}_T^{\mathrm{T}} = \begin{pmatrix} \frac{\partial L_{\mathrm{E},1}}{\partial T} & 0 & 0 \\ 0 & \frac{\partial L_{\mathrm{E},2}}{\partial T} & 0 \\ 0 & 0 & \frac{\partial L_{\mathrm{E},3}}{\partial T} \end{pmatrix} \begin{pmatrix} u_T & 0 & 0 \\ 0 & u_T & 0 \\ 0 & 0 & u_T \end{pmatrix} \begin{pmatrix} 1 & 1 & 1 \\ 1 & 1 & 1 \\ 1 & 1 & 1 \end{pmatrix} \begin{pmatrix} u_T & 0 & 0 \\ 0 & u_T & 0 \\ 0 & 0 & u_T \end{pmatrix}^{\mathrm{T}} \begin{pmatrix} \frac{\partial L_{\mathrm{E},1}}{\partial T} & 0 & 0 \\ 0 & \frac{\partial L_{\mathrm{E},2}}{\partial T} & 0 \\ 0 & 0 & \frac{\partial L_{\mathrm{E},3}}{\partial T} \end{pmatrix}^{\mathrm{T}} . \quad (3)$$

As another example, for the noise in the Earth counts, which is an effect with no channel-to-channel error correlation, for the central matrix $\boldsymbol{R}_k$ will be a diagonal matrix with 1s down the diagonal and zeroes elsewhere and the matrix $\boldsymbol{U}_k$ would be in units of Earth counts and give the uncertainty for each channel due to noise (which may be different channel to channel).

The covariance matrix for Earth radiance from channel-to-channel due to all effects is given by summing these covariance matrices, thus:

$$\boldsymbol{S}_c = \sum_k \boldsymbol{C}_k\boldsymbol{U}_k\boldsymbol{R}_k\boldsymbol{U}_k^{\mathrm{T}}\boldsymbol{C}_k^{\mathrm{T}} . \quad (4)$$

3.2.3. Error covariance between spatial pixels

When data processing requires combining data from different spatial pixels, for example to give regional spatial averages, we need to determine the error covariance between the Earth radiance values in different pixels. This can be determined in a similar manner to the error covariance between channels. For each effect, three matrices are needed for the sensitivity coefficients, standard uncertainties and correlation coefficients. These are square matrices with dimension equal to the number of pixels considered. The correlation matrix $\boldsymbol{R}_k$ will express the error correlation structure for the effect and will depend on both the correlation form, discussed above, and the separation of the pixels being considered, in each of the dimensions for correlation. So, for example, if we consider that the calibration process is averaged with a rolling average over three scanlines (the current one and one either side), then the error correlation between two pixels in the same scanline due to the calibration process is 1, and the error correlation between two pixels in neighbouring scanlines is 2/3 (two scanlines in common in averaging), and the error correlation between two pixels two scanlines apart is 1/3. This means that if we have a process that averages over nine pixels, as in Figure 1, then the error correlation, due to the uncertainty associated with that rolling-averaged calibration process, between pairs of pixels is 1, 2/3 or 1/3 depending on whether they are in the same scanlines, neighbouring scanlines or scanlines further apart.

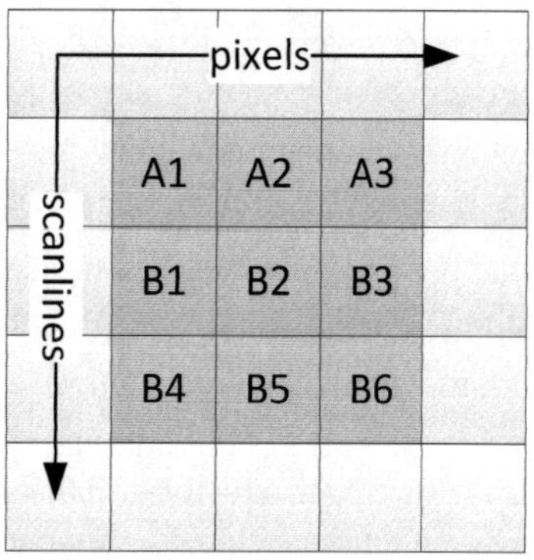

$$
\begin{array}{c} \\ \boldsymbol{R}_k = \end{array}
\begin{array}{c|ccccccccc|}
 & \mathrm{A1} & \mathrm{A2} & \mathrm{A3} & \mathrm{B1} & \mathrm{B2} & \mathrm{B3} & \mathrm{C1} & \mathrm{C2} & \mathrm{C3} \\
\mathrm{A1} & 1 & 1 & 1 & 2/3 & 2/3 & 2/3 & 1/3 & 1/3 & 1/3 \\
\mathrm{A2} & 1 & 1 & 1 & 2/3 & 2/3 & 2/3 & 1/3 & 1/3 & 1/3 \\
\mathrm{A3} & 1 & 1 & 1 & 2/3 & 2/3 & 2/3 & 1/3 & 1/3 & 1/3 \\
\mathrm{B1} & 2/3 & 2/3 & 2/3 & 1 & 1 & 1 & 2/3 & 2/3 & 2/3 \\
\mathrm{B2} & 2/3 & 2/3 & 2/3 & 1 & 1 & 1 & 2/3 & 2/3 & 2/3 \\
\mathrm{B3} & 2/3 & 2/3 & 2/3 & 1 & 1 & 1 & 2/3 & 2/3 & 2/3 \\
\mathrm{C1} & 1/3 & 1/3 & 1/3 & 2/3 & 2/3 & 2/3 & 1 & 1 & 1 \\
\mathrm{C2} & 1/3 & 1/3 & 1/3 & 2/3 & 2/3 & 2/3 & 1 & 1 & 1 \\
\mathrm{C3} & 1/3 & 1/3 & 1/3 & 2/3 & 2/3 & 2/3 & 1 & 1 & 1
\end{array}
$$

Figure 1 (left): A representation of nine pixels that are combined in a regional average, (right) the form of the correlation coefficient matrix for an effect that is based on a rolling average over 3 scanlines, e.g. from a common calibration process. The correlation coefficient matrix has a dimension 9 × 9 representing the different pixels. The error due to this calibration as a correlation of 1 for pixels within the same scanline (sharing all three calibration points), of 2/3rds for pixels in neighbouring scanlines (sharing two of three calibration points) and of 1/3rd for pixels two scanlines apart, sharing 1 of 3 calibration points.

This would be multiplied by the diagonal matrix $\boldsymbol{U}_k$ representing the uncertainties associated with the calibration in each of the nine pixels, and the diagonal matrix $\boldsymbol{C}_k$ giving the sensitivity coefficient for the Earth radiance in each of the nine pixels to the calibration. Other error effects will have a different correlation structure, and overall the error covariance for Earth radiance for the set of pixels will be obtained by adding together the error covariance matrices for each effect, thus:

$$\boldsymbol{S}_p = \sum_k \boldsymbol{C}_k \boldsymbol{U}_k \boldsymbol{R}_k \boldsymbol{U}_k^{\mathrm{T}} \boldsymbol{C}_k^{\mathrm{T}} \ . \tag{5}$$

IOP Conf. Series: Journal of Physics: Conf. Series **972** (2018) 012003 doi:10.1088/1742-6596/972/1/012003

3.2.4. Propagation to CDRs

A CDR is determined from an FCDR through a processing step that combines information from the FCDR (measured radiance values in each Earth pixel, potentially combining different spectral bands), with external information about the environmental state. While uncertainties associated with the additional information must be considered separately, the uncertainties from the FCDR can be straightforwardly propagated using the Law of Propagation of Uncertainty together with the error covariance information described above. Where a CDR is gridded in time or space (i.e. averaged over an area and/or time), the information about pixel-to-pixel error covariances must be included to obtain the correct uncertainties.

4. Harmonisation of the measurement equation

The harmonisation coefficients (the a_i in Eq.(1)) are determined through a regression analysis using sensor-to-sensor match-ups for different pairs of sensors in the series (creating a chain from the present day to the first satellite launched) and, where a suitable reference is available, from at least one sensor in that chain to a suitable independent reference, solving simultaneously the calibration parameters for all sensors in the series, connected through a set of match-ups between pairs of sensors.

The harmonisation problem is a multi-parameter, non-linear least-squares regression problem, with a very large data set (typically, hundreds of millions of match-ups), where there are error correlations between data relating to different match-ups, and where there are uncertainties for all measured variables. Solving this regression problem requires an "Errors in Variables" approach that can handle the full correlation structure and is a subject of ongoing scientific research that will be reported in a later publication.

The process of harmonisation will provide the harmonisation coefficients (the a_i) and the associated error covariance matrix; because the coefficients are determined from the same set of data, there will be an error covariance term associated with each pair of coefficients.

5. Conclusions

In this paper we have introduced some key principles for the application of metrological techniques to historical satellite sensors and the FCDRs and CDRs. Within the European project FIDUCEO we are currently developing full analyses for four different FCDRs in the visible, thermal infrared and microwave spectral regions. The process undertaken for this analysis has been described.

Acknowledgments

This work was supported by FIDUCEO, which has received funding from the European Union's Horizon 2020 Programme for Research and Innovation under Grant Agreement no. 638822.

References

[1] FIDUCEO: Fidelity and Uncertainty in Climate data records from Earth Observations; www.fiduceo.eu

[2] Merchant, CJ, Mittaz, JPD, Woolliams ER. Applying principles of metrology to historical Earth observations from satellites, in preparation, to be submitted to Metrologia 2018.

[3] Mittaz, J.P.D, Harris, A.R., Sullivan, J.T., A Physical Method for the Calibration of the AVHRR/3 Thermal IR Channels 1: The Prelaunch Calibration Data. J. Atmos. Oceanic Technol., 26, 996–1019, 2009.

[4] JCGM 100:2008 Evaluation of measurement data – Guide to the expression of uncertainty in measurement [GUM], Report, http://www.bipm.org/en/publications/guides, (2008)

IOP Conf. Series: Journal of Physics: Conf. Series **972** (2018) 012004 doi:10.1088/1742-6596/972/1/012004

Principal component analysis of reference sites used for calibration and validation of Earth observation satellites

Joaquin Campos[1], Alejandro Ferrero[1], Emma Woolliams[2], Claire Greenwell[2], Agnieszka Bialek[2], Luisa Hernanz[1] and Alicia Pons[1]

[1]Instituto de Optica (IO-CSIC), Madrid, Spain. [2]National Physical laboratory (NPL) Teddington, United Kingdom

joaquin.campos@csic.es

Abstract. Determining reflectance factor and its variability across reference sites for Earth observation satellites is a problem involving large amounts of data and measurement time. Principal component analysis (PCA) may be used to simplify this problem by reducing the size of the data and by highlighting spectral features that could be related to physical phenomena. This work presents the results obtained in applying PCA to two reference sites for calibration and validation of Earth observation satellites located at La Crau (France) and Gobabeb (Namibia), respectively.

1. Introduction

The Earth observation community requires the use of terrestrial reference sites to validate and monitor the performance of satellite sensors. Ideally these should be stable long term and if used for calibration, be traceably characterized. The Committee on Earth Observation Satellites (CEOS) has identified an initial set of reference targets for the community: Pseudo Invariant Calibration Sites (PICS) e.g. Saharan deserts, which have long term radiometric stability but are inaccessible to direct measurements by ground teams and a new network of sites which have autonomous instrumentation continually measuring the surface reflectance and atmospheric conditions, called RadCalNet. Because satellites observe with a resolution which is typically 10 m – 300 m and do not have the geometric knowledge to ensure exact co-location spatial, the spatial uniformity of such sites is critical, as is knowledge of the spectral properties of the surface.

To analyze the ground uniformity, measurements are made across the area in a sampling strategy that is consistent with the spatial resolution of the satellites. For each surface element measured we determine the reflectance factor $\rho_r(x_i,y_i,\lambda_l)$, where (x_i,y_i) are the ground coordinates of the centre of the surface element and λ_l is the wavelength that characterizes the spectral band analyzed.

Principal Component Analysis (PCA) [1] is a powerful statistical technique that can be used to identify and quantify uncorrelated contributions to the total variance of a collection of data. This method has been successfully applied to different scenarios, providing physical insight about the origin of noise contributions and extracting meaningful signals from a wide variety of situations [2, 3].

2. Principal component analysis

If the number of surface elements of a reference site is N and the number of spectral values in each spectrum is L, the total number of data points to be handled is $N \cdot L$. The PCA method produces three types of elements: N eigenvalues, γ_j, N eigenvectors, $c_{i,j}$, and N eigenspectra, H_j. The eigenvalues quantify the importance and the contribution to the total data variance of their associated eigenspectra. The eigenvectors can be seen as the coefficients of the transformation from the correlated variables given by the spectra to a new set of uncorrelated variables expressed by the eigenspectra.

Following PCA, spectral reflectance factor values can be recovered as:

$$\rho_r(x_i, y_i, \lambda_l) = \langle \rho_r(x_i, y_i) \rangle_\lambda \cdot \left[1 + \sum_{j=1}^{M} c_j(x_i, y_i) \cdot H_j(\lambda_l)\right] \qquad (1)$$

where $\langle \rho_r(x_i, y_i) \rangle_\lambda$ is the spectral average on the surface element located at (x_i, y_i) and M is the number of eigenspectra needed to reproduce the reflectance factor spectrum. Here, H_j and $c_j(x_i,y_i)$ were rescaled so that the standard deviation of H_j is 1. This way, the value of $c_j(x_i,y_i)$ quantifies the contribution of every principal component at the locations. This approach also yields the most significant results for PCA, which performs best in a null-average-value data set.

The dimensionality of the problem will be significantly reduced if pattern spectra can be found underlying the reflectance factor values, making it unnecessary to use all the eigenspectra (i.e. $M << N$). This is likely to be the case for satellite reference sites because they are chosen as areas with relatively uniform ground reflectance. The number M has to be determined by taking into account the contribution to the total variance of each eigenspectra given by γ_j and the uncertainty associated with the calibration.

3. Calibration sites studied

PCA analysis has been applied to two reference sites: La Crau (France), where 14 surface elements were selected, and Gobabeb (Namibia), where 16 surface elements were selected. La Crau and Gobabeb are two of the four prototype sites for the new CEOS Radiometric Calibration Network (RadCalNet); La Crau is an established site that has been operated by CNES since 1987 [4], while Gobabeb is a new site being developed by ESA, CNES and NPL [5]. The spatial distribution of the surface elements measured in this work are shown in figure 1, left for La Crau and right for Gobabeb. The measurements at Gobabeb are separated into clusters since the measurements were gathered during the initial characterisation of the site, meaning measurements were needed over a wide area, unlike La Crau, where this was a routine check for an established site. Different colours in the Gobabeb site image correspond to different sets of measurements, taken to represent different parts of the site.

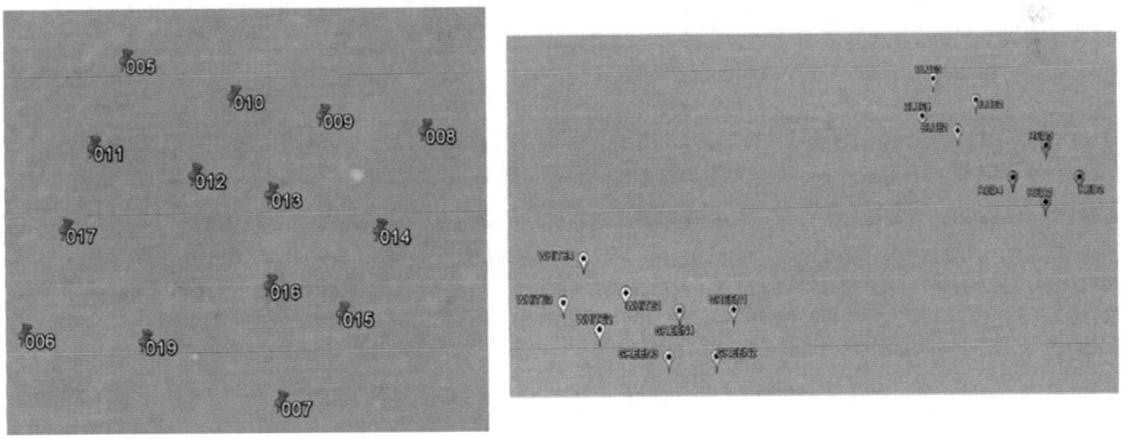

Figure 1. Spatial distribution of surface elements measured at La Crau (left-hand side) and Gobabeb (right-hand side)

A typical relative spectral distribution of the reflectance factor at a surface element of each site is shown in the Figures 2 and 3. The measured values in the noisy spectral region around 1850 nm were excluded from the analysis described in this paper (this is likely to be due to changes in the atmospheric conditions). The spectral properties are qualitatively different between the two sites. At La Crau, (figure 2) the spectral data decay smoothly from the maximum (excluding the noisy region around 1850 nm) towards both spectral ends, while at Gobabeb, the spectral reflectance is flatter over a wider spectral region, with a faster spectral decay at both spectral ends.

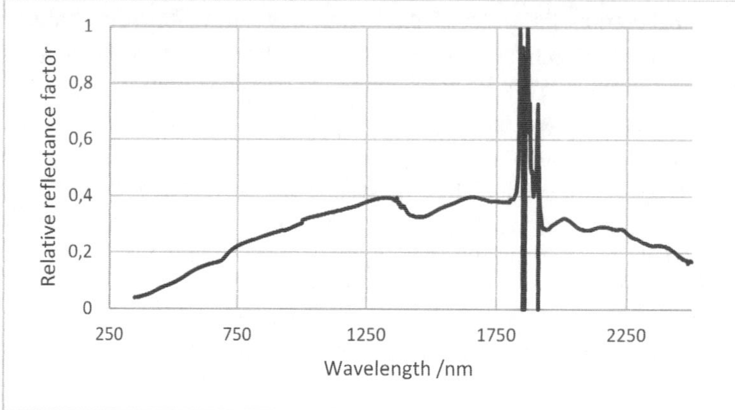

Figure 2. Typical relative spectral distribution of reflectance factor at La Crau (France)

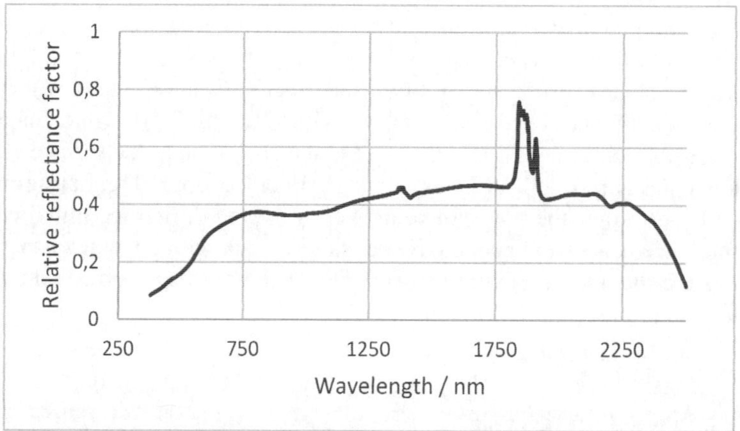

Figure 3. Typical spectral distribution of reflectance factor at Gobabeb (Namibia)

4. PCA results for reference sites

The first step of PCA analysis is to look at the eigenvalues γ_i of the uncorrelated components and calculate their contribution to the total variability of the data, because this will show how many principal components need to be considered. Table 1 shows these contributions for both calibration sites.

Table 1. Relative contribution of the eigenvalues of both calibration sites

La Crau		Gobabeb	
Eigenvalue	Contribution	Eigenvalue	Contribution
γ_1	$9.93\cdot 10^{-01}$	γ_1	$9.97\cdot 10^{-01}$
γ_2	$2.93\cdot 10^{-03}$	γ_2	$2.97\cdot 10^{-03}$
γ_3	$2.15\cdot 10^{-03}$	γ_3	$2.46\cdot 10^{-04}$
γ_4	$7.75\cdot 10^{-04}$	γ_4	$5.03\cdot 10^{-05}$
γ_5	$3.19\cdot 10^{-04}$	γ_5	$3.50\cdot 10^{-05}$
γ_6	$1.37\cdot 10^{-04}$	γ_6	$2.34\cdot 10^{-05}$
γ_7	$9.87\cdot 10^{-05}$	γ_7	$2.22\cdot 10^{-05}$
γ_8	$7.03\cdot 10^{-05}$	γ_8	$1.72\cdot 10^{-05}$

La Crau		Gobabeb	
Eigenvalue	Contribution	Eigenvalue	Contribution
γ_9	$3.75\cdot 10^{-05}$	γ_9	$1.42\cdot 10^{-05}$
γ_{10}	$2.35\cdot 10^{-05}$	γ_{10}	$7.35\cdot 10^{-06}$
γ_{11}	$2.05\cdot 10^{-05}$	γ_{11}	$6.01\cdot 10^{-06}$
γ_{12}	$1.04\cdot 10^{-05}$	γ_{12}	$4.60\cdot 10^{-06}$
γ_{13}	$6.44\cdot 10^{-06}$	γ_{13}	$2.78\cdot 10^{-06}$
γ_{14}	$5.02\cdot 10^{-06}$	γ_{14}	$2.34\cdot 10^{-06}$
		γ_{15}	$1.21\cdot 10^{-06}$
		γ_{16}	$9.68\cdot 10^{-07}$

According to Table 1, the number of eigenspectra needed to reproduce spectral reflectance factors with a residual error lower than 0.1 % (contribution lower than 10^{-3}) is four for La Crau and two for Gobabeb, rather than having a different spectrum for every point, which implies that *M* (the number of spectra needed to achieve a particular uncertainty) is significantly smaller than the original number of observations. In both sites the first principal component is dominant and other components are relatively minor variations. This is to be expected for relatively uniform sites. The Gobabeb site has less spectral variation between its surface elements than La Crau because a lower number of principal components are needed to achieve a certain residual error level.

The relevant principal components obtained for each calibration site are shown in figures 4 and 5. In both sites the main component is very similar to the typical spectra shown in figures 2 and 3, respectively, except for the peaks.

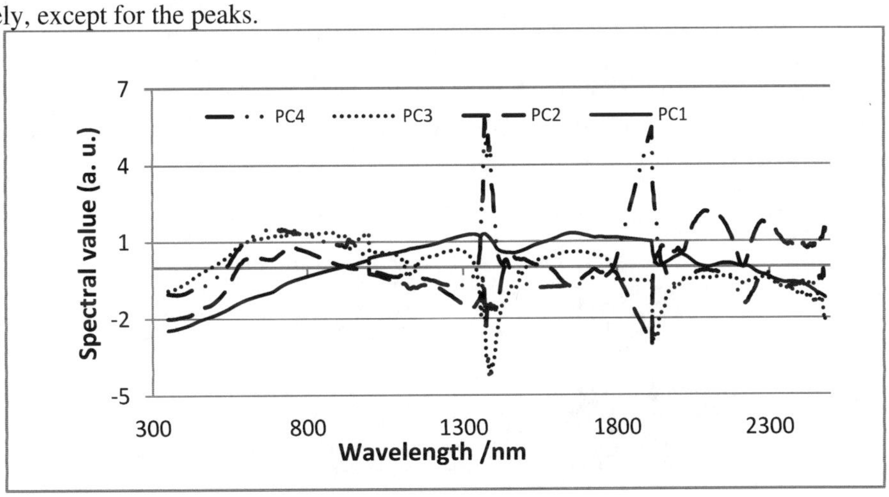

Figure 4. Relevant principal components for the calibration site at La Crau

At the La Crau site (Fig. 4) PC2, PC3 and PC4 are needed to explain the variability across the ground in the visible range, which is likely due to the change in vegetation. Furthermore, PC2 has some spectral features in the NIR that can be related to small changes in chemical composition of the ground. Finally, PC3 and PC4 also have two peaks each that could be related to changes in humidity on the ground (peak about 1400 nm due to water absorption) and the noisy peak around 1850 nm found in the raw spectra. This kind of spectral variability has an influence on the spectral behaviour of the uncertainty in the calibration of Earth observation instruments. Note that vegetation and water content may change in a surface along the time.

At Gobabeb's site (Fig. 5) the spectral variability across the ground is much less. Only 2 principal components are relevant. The ground is highly uniform in the NIR where only the peak at about 1850 nm seems to have more variability. PC2 is needed to give explanation for that variability, and for changes in wavelengths below 700 nm, which could also be due to varying vegetation levels. The variation is

IOP Conf. Series: Journal of Physics: Conf. Series **972** (2018) 012004 doi:10.1088/1742-6596/972/1/012004

smaller than for La Crau (the vegetation is sparser and drier than at La Crau), so this site has several benefits for spectral calibration compared to the La Crau site.

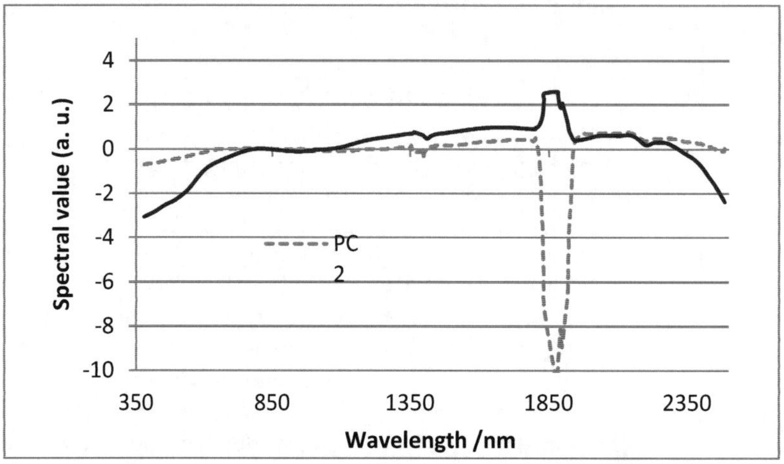

Figure 5. Relevant principal components for the calibration site at Gobabeb

Principal Component Analysis determines coefficients $c_j(x_i,y_i)$ to be used to reconstruct the reflectance factor values at each calibration position across the site. The variation of these calibration factors shows the contribution to the spectral variance of each component across the site. Fig. 6 shows $c_j(x_i,y_i)$ for each of the four principal components at each of the 14 measured locations at La Crau (spatially arranged in a grid pattern, similarly to what is shown in Figure 1, but rotated to the left); no data were available for the grid positions marked as N/A. From figure 6 it can be concluded that spatial variability at La Crau does not follow any apparent spatial pattern since no relation can be established between the coefficient values at different positions. This is consistent with the conclusion that the origin of the spectral variability is variations in vegetation levels, which is a variability on a small spatial scale on that site.

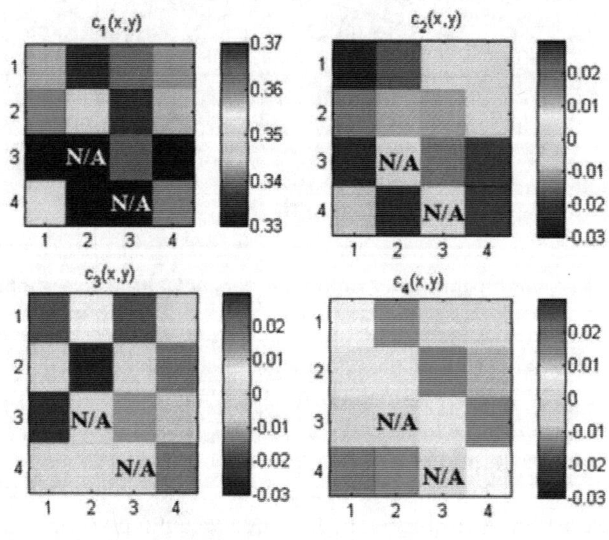

Figure 6. Spatial distribution of the coefficients for every relevant principal component at La Crau

The coefficients for the two relevant principal components at Gobabeb are shown in figure 7. In this site the coefficient of PC2 has a spatial pattern: surface elements grouped under the colour names (see Fig. 2) have the same sign. The correlation is stronger for the groups "blue", "green" and "red" (colours relate to those in Fig. 1), because the coefficient is almost the same for every surface element in the group. This shows up larger-spatial-scale variability across the site, from one group to the next, (albeit at a smaller absolute level of variation compared to that at La Crau).

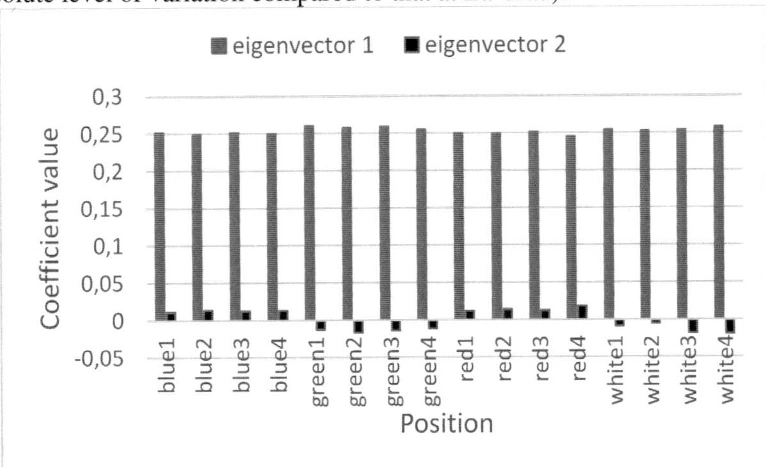

Figure 7. Coefficients for every relevant principal component at Gobabeb

5. Conclusions

PCA shows patterns hidden in the data set that may be used to analyze spectral and spatial variability between the surface elements of the reference site and to get the number of independent spectra involved in the problem. Spectral features varying across the site may be relevant to evaluate calibration uncertainty values.

PCA may help to identify physical processes influencing the reflectance factor of surface elements by associating spectral features to physical phenomena. It may help to study temporal evolution of identified phenomena if the measurements are repeated over time.

References

[1] J. L. Simonds, "Application of characteristic vector analysis to photographic and optical response data," J. Opt. Soc. Am. 53, 968–971 (1963).

[2] J. M. Lopez-Alonso, J. Alda, and E. Bernabeu, "Principal-component characterization of noise for infrared images," Appl. Opt. 41, 320–331 (2002).

[3] A. Ferrero, J. Alda, J. Campos, J. M. López-Alonso, and A. Pons, "Principal components analysis of the photoresponse nonuniformity of a matrix detector", Appl. Opt. 46, 9-17 (2007).
A. Ferrero, J. Campos, A.M. Rabal, A. Pons, M. L. Hernanz, A. Corróns, "Principal Components Analysis on the spectral Bidirectional Reflectance Distribution Function of Ceramic Colour Standards", Opt. Exp. Vol. 19, No. 20, pag. 19199-19211 (2011).

[4] A. Meygret, R. Santer, B. Berthelot, "ROSAS: A Robotic Station for Atmosphere and Surfface characterization dedicated to on-orbit calibration", Proc of SPIE, vol 8153, p 815311-1, (2011).

[5] A. Bialek, CL. Greenwell, M. Lamare, A. Meygret, S. Marcq, S. Lacherade, ER. Woolliams, B. Berthelot, M. Bouvet, M. King, C. Underwood, N. Fox, "New radiometric calibration site located at Gobabeb, Namib desert", IGARSS 2016, p. 6094-6097, doi: 10.1109/IGARSS.2016.7730592, (2016)

Acknowledgements

To the European Metrology Research Program (EMRP), supporter of this work within the joint research project ENV53 "European metrology for Earth observation and climate" (MetEOC2). The EMRP is jointly funded by the EMRP participating countries within EURAMET and the European Union. We

IOP Conf. Series: Journal of Physics: Conf. Series **972** (2018) 012004 doi:10.1088/1742-6596/972/1/012004

thank CNES for access to La Crau and their support with the measurements at both La Crau and Gobabeb. The IO-CSIC authors are also grateful to Comunidad de Madrid for funding the program SINFOTON-CM: S2013/MIT-2790.

IOP Conf. Series: Journal of Physics: Conf. Series **972** (2018) 012005 doi:10.1088/1742-6596/972/1/012005

Prototype of Cryogenic Solar Absolute Radiometer and Transfer Radiometer for On-Board Calibration of Spectral Earth Imager

L. Zajiczek[1,†], R. Winkler[1], T. Hobson[1,††], P. Green[1] and N. Fox[1]

[1]Earth Observation, Climate and Optical Group, National Physical Laboratory, Hampton Road, Teddington, Middlesex TW11 0LW United Kingdom
[†]Current address: Surgical Robot Vision Group, University College London, London WC1E 6BT United Kingdom
[††]Current address: Stephenson Institute for Renewable Energy, University of Liverpool, Liverpool L69 7ZF United Kingdom

E-mail: nigel.fox@npl.co.uk

Abstract. We describe a prototype calibration system for the Traceable Radiometry Underpinning Terrestrial- and Helio-Studies (TRUTHS) satellite mission. We outline the design and testing of key system components, including the Cryogenic Solar Absolute Radiometer (CSAR). CSAR is designed to make Total Solar Irradiance (TSI) measurements with a standard uncertainty of < 0.01 % and provide SI-traceable calibration of a hyperspectral Earth imager (EI) over the wavelength range 320 nm – 2450 nm. The EI is designed to make Solar Spectral Irradiance (SSI) and spectral radiance measurements with a standard uncertainty of < 0.3 %.

1. Introduction

The Traceable Radiometry Underpinning Terrestrial- and Helio-Studies (TRUTHS) satellite mission intends to provide SI-traceable measurements of spectral Earth radiance and Solar Spectral Irradiance (SSI) within a standard uncertainty of < 0.3 % and Total Solar Irradiance (TSI) within a standard uncertainty of < 0.01 % [1]. The unprecedented accuracy of the mission's measurement capability would not only provide benchmark climate monitoring but also offer reference calibration of other Earth observation satellites. Measurement traceability is provided by an on-board Earth imager calibration system anchored to the SI through the Cryogenic Solar Absolute Radiometer (CSAR). We present a flight-representative prototype of CSAR cooled by Airbus' High Performance Stirling Cooler (HPSC) and a breadboard calibration system demonstrating a technology readiness level (TRL) of 5 – 6. We also present an engineering assessment of the feasibility and reliability of a space-qualified version of the TRUTHS payload.

2. Design of Payload Instruments

The payload of TRUTHS consists of three main components: an on-board primary standard CSAR, a hyperspectral Earth imager (EI) and a calibration system that anchors the EI measurements of spectral radiance to CSAR via a transfer radiometer (TR). A suite of power- stabilized single-wavelength laser diodes over the range of 320 nm to 2450 nm serve as anchor points in the calibration of the EI via the TR and an illuminated Spectralon diffuser target.

Laser light is collimated and directed into different instruments as needed during the calibration chain via a rotating roof prism arm (RPA). CSAR provides both TSI measurements and an absolute measurement of each laser diode's optical power, calibrating the response of the TR at that wavelength. The diffuser is then illuminated by the laser light and the calibrated TR measures the radiance of the diffuser simultaneously with the EI, calibrating its response. Finally a white light source (incandescent current/voltage stabilized lamp, although as this is only used in a relative mode anchored spectrally this is less critical than would be the case if used as an absolute source) is used to illuminate the diffuser and the calibration of the EI is interpolated between the laser diode wavelengths over the entire spectral range. Lamps of this nature have been flown in space before, are widely used terrestrially and by their physics have relatively smoothly varying spectral irradiance. However, both in space and on ground the absolute level of the output and its absolute spectral shape is subject to variation, with the latter dependent on its nominal 'Planckian' temperature/emissivity although the latter will only result in a spectrally smooth and relatively small variation which can be anchored by fixed spectral (laser diode) calibration points. This type of lamp is A schematic of all major components in the payload is shown in Fig. 1 with a flow chart of the calibration chain shown in Fig. 2.

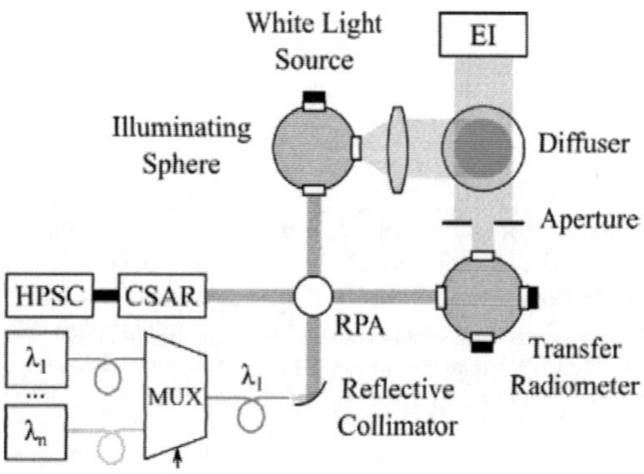

Figure 1. Schematic of calibration system.

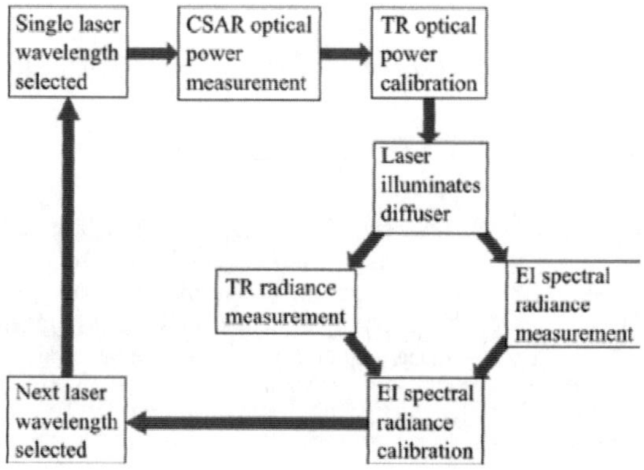

Figure 2. Flow chart showing calibration chain of EI at each laser wavelength.

2.1. *Cryogenic Solar Absolute Radiometer*

The design of the onboard cryogenic radiometer is based on the terrestrial Cryogenic Solar Absolute Radiometer located at the World Radiation Centre in Davos, Switzerland. This instrument is used to measure direct solar radiation and as it is directly traceable to the SI, it is expected to replace the existing World Standard Group of independently-designed instruments in the near future [2]. Terrestrial cryogenic radiometers including CSAR are operated at temperatures between 10 K – 20 K which would be difficult to achieve with existing space qualified cryocoolers [3]. Airbus' recently optimized High Performance Stirling Cooler (HPSC), based on the high-heritage Astrium 50-80K cryocooler, can readily achieve a 45 K cold tip temperature with a 1 W heat load. The terrestrial CSAR design was significantly modified to reduce its volume, mass, complexity, and crucially, surface area with the aim of reducing the heat load of the instrument to 1 W as shown in Fig 3.

Initial testing of the thermal response of the redesigned cavity at an expected operating temperature of 60 K yielded a time constant of $\tau = 165$ seconds and a sensitivity of 0.0004 %, well within the 0.01 % requirements of the mission. Rather than measuring the absolute change in temperature due to an introduction of optical radiation requiring several multiples of τ to stabilize, the cavity temperature was actively maintained by a PID loop. The absolute difference in electrical power through the cavity's resistive heater before and after the introduction of optical radiation is equivalent to the optical power of the radiation source. This introduces additional instability due to the uncertainties of the current source output and control loop, but the reduction in measurement time allows for additional integration of cavity temperature measurements to measure the electrical power within the required uncertainty.

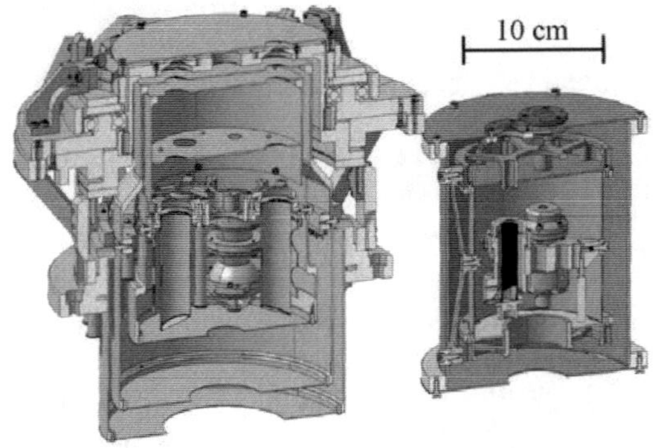

Figure 3. Comparison of terrestrial CSAR design (left) and modified space CSAR design (right).

2.2. Calibration Breadboard
The transfer radiometer (TR) is an intermediate radiance-measuring instrument whose response coefficient is calibrated by CSAR at each laser wavelength. The large calibration range necessitates three detectors: a silicon (Si) photodiode for the ultraviolet-visible-near infrared portion of the range, a thermo-electrically cooled indium gallium arsenide (InGaAs) photodiode for the near and shortwave infrared and an extended range InGaAs detector for full calibration to 2450 nm. To ensure the radiance incident on the detectors was as uniform as possible for both viewing conditions and at all wavelengths, the TR was designed as a 50.8 mm diameter Spectralon integrating sphere with a single input port and three detector ports. The RPA is rotated into position between the TR's input port and its radiance-defining external aperture for the response calibration. The diffuser target is illuminated via another 50.8 mm diameter Spectralon integrating sphere with two input ports and one output port. The RPA directs laser light into its first input port, and the output port radiance is collimated and directed onto the diffuser via a periscope as the integrating sphere is situated at a different height than the diffuser disk. Two different collimating configurations were investigated and are presented in Section 3.3. The TR measures the radiance of the illuminated diffuser target at a 45 degree viewing angle and with a similar field of view to the EI. The illuminating sphere's second input port contains a white light source which is used for interpolating the EI calibration between the anchor wavelengths.

3. Prototype Assembly and Testing
A flight representative cryostat was thermally modelled and procured by Airbus to house the CSAR instrument and the HPSC. CSAR's thermal performance was tested separately before being mounted to a larger vacuum tank housing the calibration system prototype; the assembled prototype inside the tank is shown in Fig. 4. As a first test, both a trap detector with an external aperture acting as an EI analogue and the TR viewed the illuminated diffuser disk at a 45 degree angle and its radiance was measured. A flat-panel light source was then illuminated with the same laser and placed directly in front of both instruments (approximately 5 cm away); its radiance was measured, and the ratios of the EI and TR radiance measurements for both configurations agreed to within 0.08 %.

3.1. Thermal Testing of CSAR
The HPSC achieved a minimum CSAR reference block temperature of 55 K which was in line with performance expectations. The measured cavity time constant and temperature sensitivity at a cavity temperature of 60.12 K were consistent with the thermal model presented in Section 2.1: $\tau = 217$ seconds and a better sensitivity than expected (0.09 K/mW). The cavity was then stabilized 15 mW or 1.35 K above 60.12 K, with 15 mW being in excess of a typical TSI measurement. Typical excursions of the cavity temperature from its setpoint due to optical

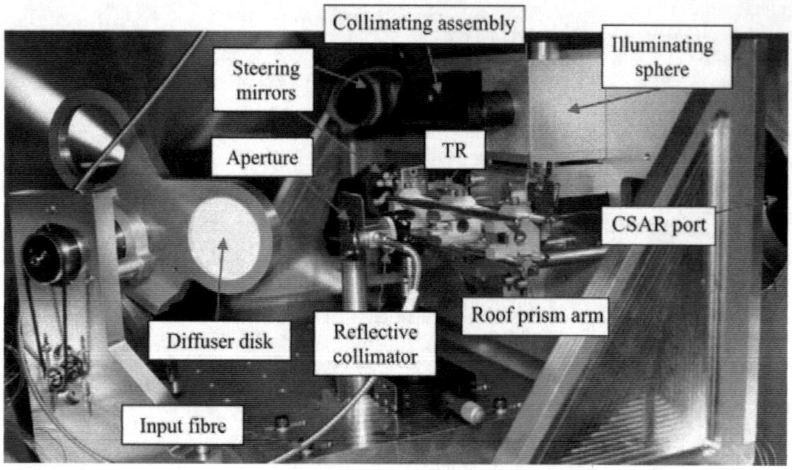

Figure 4. Assembled calibration system breadboard inside vacuum tank.

radiation heating the cavity would be on the order of 1 % – 5 %. The stability of the cavity temperature after such an excursion is nominally required to be within 0.01 % of this 15 mW uplift, and while the PID control loop was not able to achieve this stability criterion by itself, the temperature quickly stabilized to within 0.2 %. The standard deviation of the cavity temperature measurements was just under 4 times the 0.01 % requirement, meaning an additional integration time of 10 seconds or 20 samples at 2 Hz would be sufficient. For a worst case scenario of a 5 % excursion, a cavity temperature stabilization achieving the required stability criterion took 38 seconds in total. The time needed to make a cavity power measurement would also add a finite but similarly small amount of time for an accurate optical power measurement.

3.2. *End-to-End Calibration* In Situ

The CSAR cryostat was mounted to the testing vacuum tank's viewing port and an end-to-end calibration was carried out at a single wavelength (785 nm). Once the cavity temperature was stabilized, CSAR measured an input optical power of 1.266 mW 0.001 mW with a standard uncertainty of 0.08 %. Laser light was then coupled directly into the TR and the Si detector voltage was measured at 7.07 V 0.04 V (with 105 amplifier gain), yielding calibration coefficient of 0.179 mW/V 0.006 mW/V with a standard uncertainty of 0.3 %. The illuminated diffuser radiance was measured by the TR's Si detector to be 0.417 V ± 0.004 V (with 109 amplifier gain). The EI analogue measured a signal of 0.601 V ±0.002 V, which is consistent with its larger field of view compared to the TR. Given the known diameters of the TR input aperture and external aperture as well as their separating distance, this gives a radiance of L_{diff} = 0.025 W m−2 sr−1. This value was validated by estimating the input power to the illuminating sphere that would generate an irradiance of this magnitude on the diffuser, assuming that radiance and irradiance are roughly equivalent for a Lambertian source:

$$P_{input} = \frac{L_{diff} A_{spot} \pi A_{sphere}}{M A_{port} \theta}$$

The collimation angle can be estimated from the collimating lens' clear aperture of 45 mm and focal length of 170 mm as $\theta = 0.77$ rad. Assuming a Spectralon reflectance of 95 % and a port fraction of 5 % gives a sphere multiplier M 10. For a diffuser spot 40 mm in diameter, a sphere diameter of 50.8 mm and an output port 11 mm in diameter, the input power required to generate a diffuser irradiance of 0.025 W/m2 is approximately 1.3 mW, which is in good agreement with the value measured by CSAR.

3.3. *Uniformity of Illumination*

Two critical aspects to generating uniform diffuser illumination are the collimation of the illuminating sphere output and laser speckle. The use of single-wavelength (and thus highly coherent) laser sources resulted in observed speckle patterns on the diffuser. The use of a multimode optical fibre to deliver all wavelengths of the laser diode array allowed for vibration of the fibre to mix the spatial modes, reducing the coherence and the speckle inhomogeneity. Adding a separate de-speckling element would introduce unwanted complexity, but fortunately the HPSC's compressors continuously vibrate at a frequency of 36 Hz. The multimode fibre was wrapped around a vibrating element and the illuminated diffuser was imaged with a 12-bit CCD camera. The illumination with and without 36 Hz vibration are shown in Fig. 5 using radiation from a 530 nm laser diode. The speckle generated with no vibration showed up to 20 % variation in signal (ratio of RMS to mean), while the maximum variation was 1.15 % with vibration (note that some of this residual noise is due to the performance of the CCD camera, We expect that with optimized winding of the fibre maximizing spatial mode mixing the non-uniformity of the speckle pattern will be further reduced (subsequent as yet to be published measurements using a different facility at NPL have indicated that the effect of speckle on the calibration of a spectrometer can be reduced to <0.1%). It should be noted that whilst the size of the speckle features will change as a function of wavelength, the use of the mode scrambling in the optical fibre and the various independent diffusing elements serves to minimize the effect of this spectral dependency on the overall calibration uncertainty. Measurements are still to be performed to quantify any residual spectral effects.

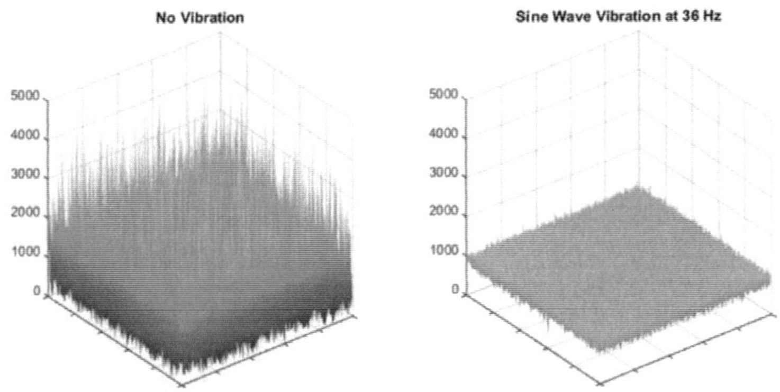

Figure 5. (left) Image of speckle pattern generated on diffuser. (right) Image of diffuser illumination with fibre vibrated at 36 Hz.

The uniformity of the diffuser radiance with despeckling was directly measured with a trap detector and external aperture having a deliberately small field of view to avoid averaging of any non-uniformity. The variation in radiance was 30 % for a change in lateral position of 10 cm, and the diffuser radiance measured with a field of view similar to the TR/EI at different distances and positions showed variations of up to a maximum of 0.25 %. The original design used a single fluorite lens to collimate the output light of the illumination sphere, and this configuration was generating a divergent collimated beam and introducing aberrations to the wavefront, resulting in a non- uniform illumination profile. A second optical element was then introduced immediately after the output port of the sphere in a Galilean beam expander configuration, and the non-uniformity was reduced from 30 % to approximately 1 %. It is expected that further improvement in uniformity can be achieved with a lens configuration that balances aberrations.

4. Reliability and Feasibility Analysis

The reliability of the TRUTHS payload was calculated as R = 0.971 and 0.955 and for a 5 year and 7 year mission respectively, based on redundancy provided by backup measurement cavities and laser diodes. No critical areas for development were identified. The estimated reliability figures were considered adequate for the level of complexity involved. The main reliability drivers were identified as the HPSC drive electronics, displacer and compressors, the TR detectors and the on-board computer controlling CSAR. It was determined that the TRUTHS control electronics can be constructed with space-qualified components meeting all performance requirements. The uncertainty budgets of the lab-based calibration system and the predicted final design for all wavelengths from around 400 to 2000 nm are given in Table 1, outside of this spectral region the uncertainty may increase a little due to performance of sources and detectors.

Table 1. Uncertainty budget for calibration system (400 to 2300 nm).

Component	Prototype	Final Design
Laser diodes	0.3 %	0.07 %
CSAR	0.08 %	0.02 %
TR	0.3 %	0.08 %
Aperture geometry	0.17 %	0.02 %
ghIllumination uniformity	0.25 %	0.02%
Total	0.4 %	0.1 %

The uncertainties for the laser sources were provided via testing through the MetEOC-2 project, and the space-qualified laser diodes are known to have improved power stability. The higher uncertainties for the CSAR and TR measurements were due to a loss of 75 % of the available laser power; the necessary addition of a port aligner between the CSAR cryostat and the testing vacuum tank increased the distance between the reflective collimator and the CSAR port, and divergence of the collimated beam necessitated using a smaller core fibre to allow the entire beam to pass through CSAR's input aperture. This constraint will not be present in the final design and as such the signal-to-noise ratio will improve by at least a factor of 4. Finally, the aperture geometry uncertainty can be reduced through pre-flight calibration and the uniformity of the diffuser is expected to improve with optimized despeckling of the fibre and improved design of the collimation optics.

5. Summary and Future Work

A lab-based prototype of the calibration system was used in vacuum to calibrate an EI analogue instrument via the redesigned space CSAR. Testing indicates that the uncertainty for TSI measurements is within 0.01 %, and the uncertainties for SSI and spectral radiance are expected to be within 0.1 %. Analysis also suggests that a space-qualified system can be built to within the specifications and achieves sufficient reliability criteria. Some further work is required before a space-qualified design can be achieved, including the following:

- Procure and test a passive laser conjoiner, e.g. a waveguide rod or hexagonal fibre.
- Optimize diffuser illumination optics and fibre winding to improve diffuser uniformity.
- Introduce an imaging spectrometer as a closer Earth imager analogue.

Acknowledgments

This work was funded by the UK space agency through its Centre for Earth Observation Instrumentation (CEOI) and the Metrology for Earth Observation and Climate project (MetEOC-2), grant number ENV55 532 within the European Metrology Research Programme (EMRP). The EMRP 533 is jointly funded by the EMRP participating countries within EURAMET.

References

[1] Fox N, Kaiser-Weiss A, Schmutz W, Thome K, Young D, Wielicki, B, Winkler R and Woolliams E 2011 Accurate radiometry from space: an essential tool for climate studies **369** *Phil. Trans. R. Soc. A* 4028–4063
[2] Walter B, Finsterle W, Mingard N and Soder R 2015 Cryogenic absolute solar radiometer (CSAR) *Physikalisch-Metorologisches Observatorium Davos/World Radiation Center Annual Report* 11
[3] R Radebaugh 2009 Cryocoolers: the state of the art and recent developments *J. Phys. Condens. Matt.* **21** 16

IOP Conf. Series: Journal of Physics: Conf. Series **972** (2018) 012006 doi:10.1088/1742-6596/972/1/012006

Traceability of the Network for Detection of the Mesospheric Change (NDMC) to radiometric standards via a Near Infrared Filter Radiometer

Steven van den Berg[1,3], Paul Dekker[1], Max Reiniger[2], Berndt Gutschwager[2], Christian Monte[2] and Jörg Hollandt[2]

[1] VSL, Thijsseweg 11, 2629 JA Delft, The Netherlands

[2] Physikalisch-Technische Bundesanstalt (PTB), WG 7.32 Infrared Radiation Thermometry, Abbestrasse 2 – 12, D-10587 Berlin, Germany

[3] Corresponding e-mail address: svdberg@vsl.nl

Abstract. We discuss a route to provide traceability to spectroradiometers that are operated within the network for detection of the mesospheric change. The spectrometers measure spectral radiance at low light levels (350 W m^{-2} sr^{-1} m^{-1}) at 1.55 μm. The traceability chain is based on a travelling reference source that is calibrated against a black body via a near infrared filter radiometer. In this paper the requirements, design and performance of the filter radiometer are discussed. It is demonstrated that the filter radiometer is suitable for transferring the scale at very low radiance levels, providing SI traceability to the traveling reference radiance source within the required measurement uncertainty of 1 %.

1. Introduction

The temperature of the mesopause, an atmospheric layer at an altitude of about 87 km, is one of the measures for the global temperature and has therefore been identified as an Essential Climate Variable (ECV) [1]. Ground-based mesopause measurements are performed within the Network for Detection of Mesospheric Change (NDMC), which comprises 55 stations. Within this world-wide network Ground-based Infrared P-branch Spectrometers (GRIPS) are used to investigate mesopause properties, such as composition and temperature. The GRIPS instruments observe air glow emissions from excited OH* and O* radicals and excited O_2. The mesopause temperature is derived from the relative intensities of rotational P-branch transitions in the Meinel band around 1.55 μm [2]-[4]. In order to determine this temperature with sufficiently low uncertainty, to allow for the measurement of temperature changes at the level of 1 K on a decade timescale, SI traceability to radiometric standards is required. Here, a very low radiance level of about 350 W m^{-2} sr^{-1} m^{-1} needs to be measured, since this is the typical radiance level of OH* radicals in the relevant wavelength range of 1500 nm to 1600 nm.

We explore a traceability route based on a black body Travelling Radiance Source (TRSO), which is used as a reference for the GRIPS instruments. The TRSO is calibrated against a black body primary reference via a Near-InfraRed Transfer Radiometer (NIRTR). In this paper we focus on the development of the NIRTR designed for measuring low radiance levels for scale transfer within a 1 % level of uncertainty.

2. Near infrared filter radiometer requirements and design

The method developed to provide SI traceability to GRIPS instruments is to expose the spectroradiometer to a known radiance level from a reference source (TRSO). The reference radiance source is a Planckian black body, in this case a transportable plate radiator, which is suitable for on-site measurements and sufficiently large to cover the field of view of the GRIPS detector (about 12°). The plate radiator is calibrated against a reference black body, using the filter radiometer (NIRTR) to transfer the scale. The traceability route is schematically depicted in Figure 1.

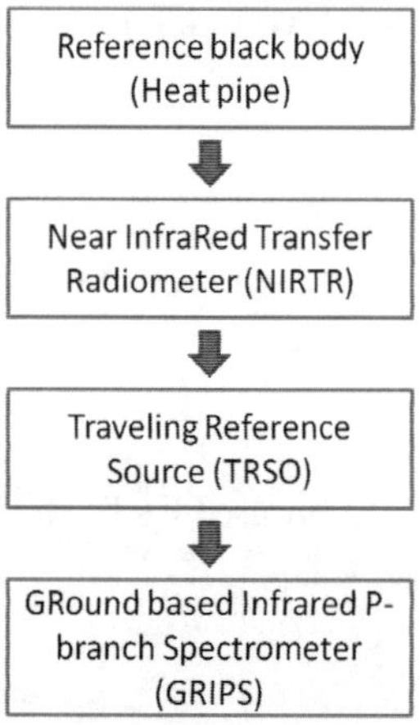

Figure 1. Traceability chain from primary black body to GRIPS instrument

2.1. Requirements

The requirements for the near-infrared transfer radiometer are based on the radiance level to be measured and on the measurement range and view angle (12°) of the GRIPS instrument. Furthermore, the NIRTR should be transportable and it should be possible to put it close to a black body radiator without affecting the NIRTR performance. To generate a typical radiance level of 350 W m^{-2} sr^{-1} m^{-1}, the reference black body and the TRSO will operate at a temperature of about 120 °C. The required measurement uncertainty on the radiance is estimated to be 1 %. The spectral range should cover the Meinel bands around 1.55 µm.

2.2. Optical design

The optical design of the NIRTR has been developed in Zemax and is shown in Figure 2 . The light is collected with an objective lens with a diameter of 50 mm and a focal length of 80 mm. The field stop, having a diameter of 4 mm, is inserted to restrict the image size on the detector. The field stop is imaged 1 to 1 on the photodiode surface with 2 lenses of 30 mm focal length. The photodiode has a sensitive area with a diameter of 5 mm and is thus underfilled. The field of view of the system is limited by a Lyot stop (16 mm) that is inserted between the two collimating lenses. The Lyot stop is inserted such that size-of-source effects are minimized [5,6]. The filter is also inserted in the collimated beam between the lenses. The filter transmission is centered around 1.55 µm, with a bandwidth of 65 nm. As a detector a thermo-electrically cooled InGaAs detector (Hamamatsu G112180) has been selected. All components are mounted along a common optical axis in a cage system consisting of four graphite epoxy rods, which minimize effects of thermal expansion. The whole instrument is shielded with a light-tight insulated housing. The front panel and bottom plate are temperature stabilized by an embedded water cooling system. A shutter is mounted in front of the objective lens to perform dark measurements.

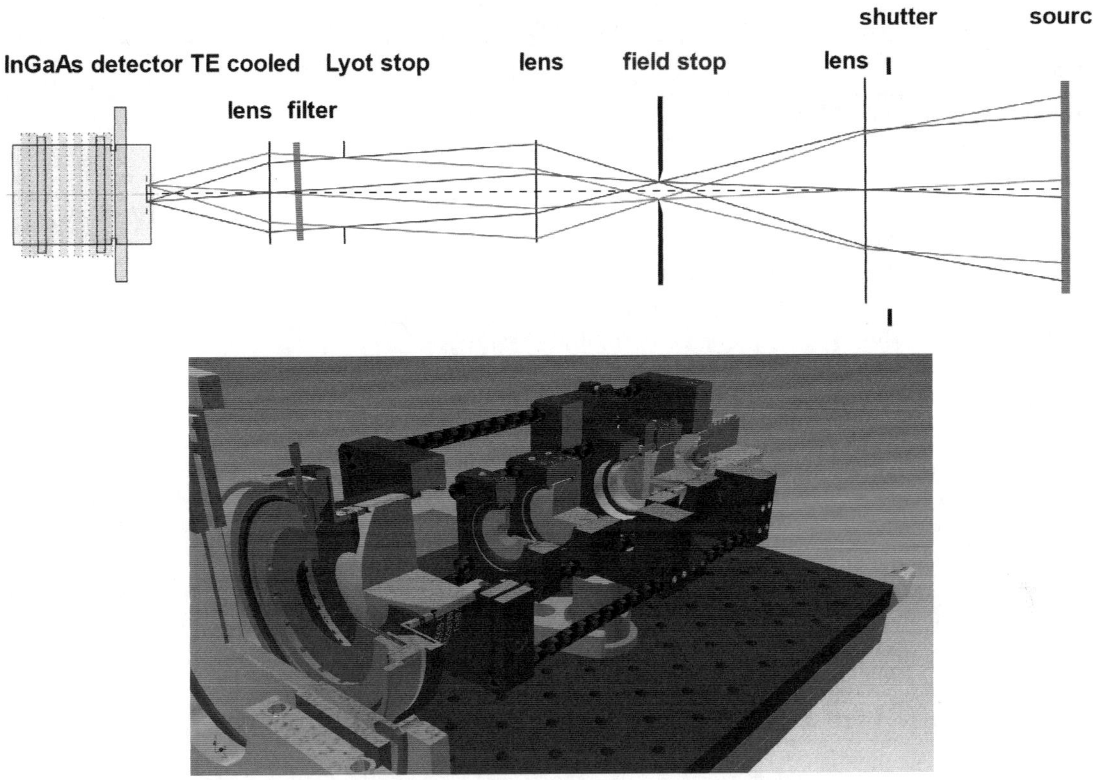

Figure 2. Design of the near-infrared filter radiometer. Top: ray tracing picture. Bottom: opto-mechanical design.

2.3. *Signal and noise levels*

Based on the design, the signal level has been estimated for a black body at a temperature of 120 °C. The signal *S(T)* of the NIRTR as a function of temperature *T* is:

$$S(T) = \int_{\lambda_1}^{\lambda_2} G \cdot R(\lambda) \cdot \tau(\lambda) \cdot \Omega \cdot A \cdot L(\lambda, T) d\lambda \,, \tag{2.1}$$

where is the G is the gain of the amplifier, $R(\lambda)$ is the spectral responsivity of the detector, $\tau(\lambda)$ is the transmittance of the interference filter, $L(T,\lambda)$ is the radiance of a black body at a temperature T, Ω is the solid viewing angle of the detector and A is the illuminated area of the detector. The lens transmission and geometrical parameters (solid viewing angle of the detector, illuminated detector area) have not been measured, but are based on the design. The detector absolute spectral responsivity and the filter transmission (including angular dependence) have been calibrated by PTB and are used here to estimate the expected signal level. For the calibration of a radiance source like the TRSO with the NIRTR, only the relative spectral dependence of responsivity and transmittance needs to be considered. Note that equation 2.1 is only used to estimate the signal level. The calibration of the NIRTR as a whole will take place against a reference black body.

Based on the design parameters as summarized in Table 1, the current generated by the photodiode is estimated to be 1.3×10^{-10} A for a radiance level of 350 W m^{-2} sr^{-1} m^{-1}. With a gain of 10^8 selected at the trans-impedance amplifier, this leads to a signal of 1.3×10^{-2} V.

Table 1. Overview of input quantities for signal level estimation

Quantity	Value
Illuminated detector area A	$\pi \times (2\ \text{mm})^2$
Solid viewing angle at detector	0.31 sr
Filter transmission	Centered around 1550 nm, width 65 nm, based on measurement data of PTB
Transmission of the system (taking into account losses due to reflection from the interfaces)	0.985
Black body emissivity	0.95
Black body temperature	120 °C
Photodiode responsivity	1.1 A/W at 1550 nm
Gain setting DLPCA-200 G	10^8

To estimate the expected noise level, the following noise contributions have been considered: the Noise Equivalent Power (NEP) generated by the photodiode, the equivalent input noise current of the current amplifier and the noise contribution of the nanoVolt meter. The NEP power of the G12180 photodiode is specified as 1.5×10^{-14} W/√Hz. The bandwidth has been limited to 10 Hz, as set on the transimpedance amplifier. In Table 2 the input parameters for the noise estimation are summarized. A total noise equivalent output voltage of 6.1 μV is expected. At an estimated signal level of 1.3×10^{-2} V, this corresponds to a signal-to-noise ratio of 0.05 %.

Table 2. Overview of input quantities for noise level estimation

Quantity	Value
NEP G12180, in 10 Hz bandwidth,	4.7×10^{-14} W
Corresponding voltage noise	4.3×10^{-6} V
Equivalent input noise current DLPCA-200	4.1×10^{-6} V
Noise contribution HP34420A @ 100×10^{-6} V/V	1.3×10^{-6} V
Total noise equivalent output voltage of system	$\mathbf{6.1\times10^{-6}}$ **V**

3. Performance and measurement results

The signal level and size-of-source effects (SSE) have been measured by exposing the NIRTR instrument to the radiance from a primary black body (water heat pipe with 60 mm aperture diameter) with a set temperature of 120 °C. SSE have been determined by measuring the signal level as a function of distance to the black body aperture. Over a distance ranging from the closest position possible (0 mm) to 140 mm, corresponding to opening angles from 61° to 18°, the signal change is within the requirements, having a standard deviation of 0.2 % (see Figure 3). For larger distances (opening angle below 18°) the NIRTR signal drops because it starts looking outside the black body aperture. This measurement illustrates the well-defined field of view of the NIRTR. Currently SSE is not expected to be the dominant contribution in the calibration of a GRIPS instrument. However, further reduction of SSE could be obtained by matching the opening angles at which the calibration takes place of NIRTR and, subsequently, the TRSO and the GRIPS instrument.

The NIRTR has been calibrated against the water heat pipe black body at PTB. From the measurement results shown in Figure 4 (left), which are obtained over 3 days of measurements, good reproducibility is observed. The noise equivalent temperature difference (NETD) based on stability measurements over 3 days is 23 mK at 120 °C (1 s averaging time). This corresponds to an uncertainty contribution of 0.14 % for the measured radiance. Based on the reproducibility of this calibration and the uncertainty resulting from SSE, the uncertainty on the calibration of the NIRTR is estimated to be 0.5 % (expanded uncertainty, coverage factor k=2).

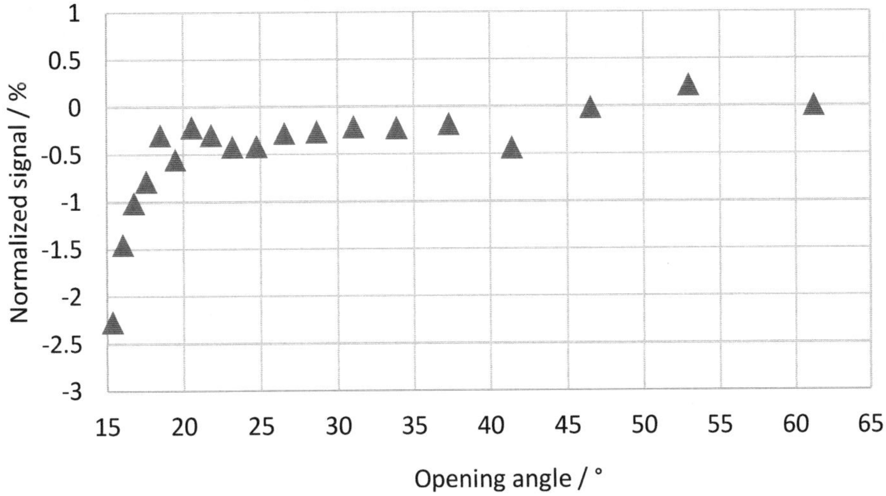

Figure 3. Size-of-source measurement: relative signal level as a function of opening angle. The standard deviation based on the range 18° to 61°, equals 0.2 %.

Subsequently, the radiance from the TRSO has been measured by the NIRTR. The difference between the set temperature and the measured radiation temperature, as based on the radiance measurement, has been plotted against the set temperature in Figure 4 (right). The difference between radiation temperature as measured in the 1.5 µm to 1.6 µm range and the set temperature results from a higher emissivity in the 1.5 µm to 1.6 µm compared to the emissivity in the 8 µm to 14 µm range (the TRSO has a factory calibration for radiation temperatures in the 8 µm to 14 µm range). In four characterization campaigns over one year the NIRTR was successfully applied to verify the long-term stability of the TRSO (Figure 4 right) and as a next step the calibrated TRSO is available to providing traceability to the GRIPS instruments.

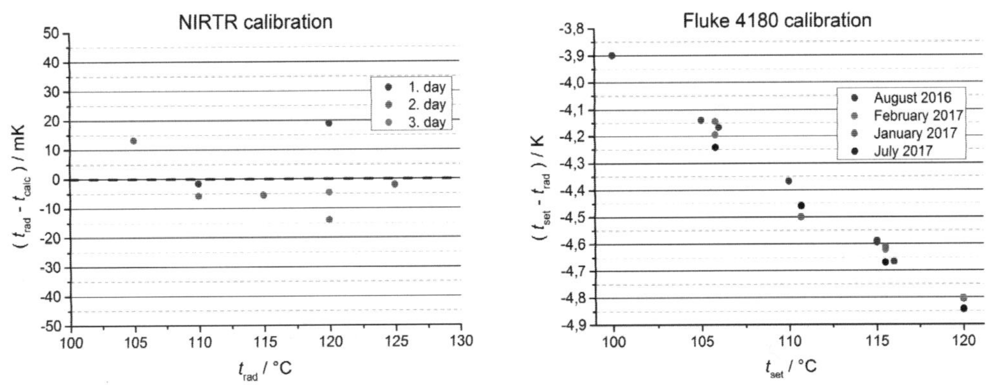

Figure 4. Left: stability of the filter radiometer expressed in mK, as measured over 3 consecutive days and for various temperatures of the water heat pipe. The noise equivalent temperature difference derived from these data is 23 mK. Right: calibration of the TRSO (Fluke 4180), temperature difference between set temperature and measured radiation temperature with the NIRTR. The relatively large difference results from the fact that the TRSO has a factory calibration for application in the 8 µm to 14 µm range.

4. Conclusions

We have presented a traceability chain for calibration of Ground-based Infrared P-branch Spectrometers, based on a travelling radiance source, that obtains its scale from a reference black body via a transfer filter radiometer. In particular, the design and performance of the NIRTR has been discussed and investigated, showing that the stability and size-of-source effects are sufficiently low to achieve a measurement uncertainty below 1 % for the radiance calibration of a black body radiator.

Acknowledgments

We acknowledge Carsten Schmidt, Sabine Wüst and Michael Bittner from DLR, who provided the input requirements on sensitivity and measurement uncertainty as needed for the calibration of a GRIPS instrument. We also acknowledge Maurice A. Heemskerk for his contribution to the mechanical design of the NIRTR.

This work was funded by the European Metrology Research Program (EMRP), Project ENV53, 'Metrology for Earth Observation and Climate 2'. The EMRP is jointly funded by the EMRP participating countries within EURAMET and the European Union.

References

[1] http://www.wmo.int/pages/prog/gcos/index.php?name=EssentialClimateVariables

[2] Bittner M, Höppner K, Pilger C and Schmidt C 2010 *Nat. Hazards Earth Syst. Sci.* **10** 1431-42

[3] Pertsev N and Perminov V 2008 *Ann. Geophys.* **26** 1049-56

[4] Meinel A B 1950 OH emission bands in the spectrum of the night sky, Part Ii, *Astrophys. Jl.*, **112**, 120-30

[5] Yoon H W, Allen D W and Saunders R D 2005 Methods to reduce the size-of-source effect in radiometers *Metrologia* **42** 89-96

[6] Dury M R, Arneil T C, Machin G and Goodman T M 2014 Size-of-source effect sensitivities in radiometers *Int. J. Thermophys.* **35**, 1391-1400

IOP Conf. Series: Journal of Physics: Conf. Series **972** (2018) 012007 doi:10.1088/1742-6596/972/1/012007

Characterisation of a new carbon nanotube detector coating for solar absolute radiometers

A Remesal Oliva[1], W Finsterle[1], B Walter[1], and W Schmutz[1]

[1] Physikalisch-Meteorologisches Observatorium Davos and World Radiation Center (PMOD/WRC), Dorfstrasse 33, 7260 Davos Dorf, Switzerland

alberto.remesal@pmodwrc.ch

Abstract. A new sprayable carbon nanotube coating for bolometric detectors aims to increase the absorptance compared to regular space qualified black paints. In collaboration with the National Institute of Standards and Technology (NIST), we have characterized the optical properties and mechanical and thermal stability of the carbon nanotube coating inside conical shaped cavity detectors.

1. Introduction

We characterized three cavity detectors coated with spray-on carbon nanotubes by Surrey NanoSystems® [1]. The cavity detectors are supposed to be used in the Digital Absolute Radiometer (DARA) developed and built for the PROBA-3 satellite. The cavities are made of silver with a gold coating layer on the exterior face, to suppress radiative heat exchange with the surroundings (e.g. [2]).

The detectors were evaluated under two different experiments. The first cavity was exposed to a high-power UV laser to study the accelerated ageing of the carbon nanotubes under optical conditions similar to 1 year of Total Solar Irradiance (TSI) space measurements. The second cavity was tested in a thermal cycling test chamber simulating thermal variations onboard a satellite. The third one served as a reference and as such was not subjected to any test. We studied the reflectance of the detectors and evaluated the coating qualitatively by Scanning Electron Microscopy (SEM) before and after the tests to evaluate the performance of the coating and to identify differences in the absorptance and in the structure of the carbon nanotubes.

2. Reflectance measurements

Reflectance maps of the cavities were produced in the optics laboratory at PMOD/WRC (Figure 1) using a setup with an integrating sphere as described in [2, 3]. Two laser wavelengths (633 nm and 375 nm) were used to study the spectral response of the coating. Both lasers have a stable output power of approximately 5 mW.

The same cavities were subjected to reflectance measurements in the Laboratory for Atmospheric and Space Physics (LASP) (Figure 2) facilities at 406 nm, 520 nm, 660 nm, 830 nm, 1310 nm, and 1625 nm.

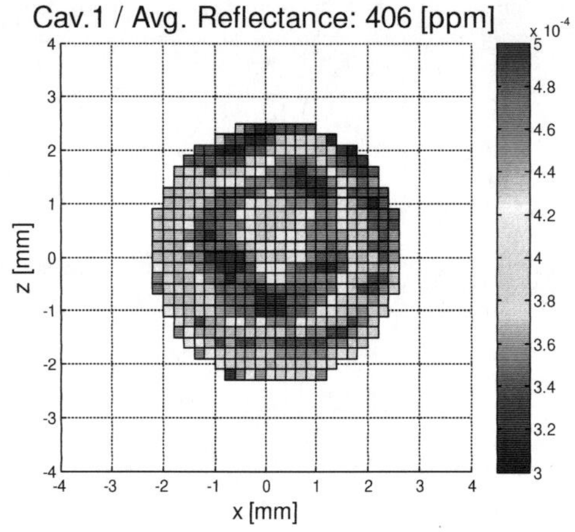

Figure 1. - Example of a measurement done at PMOD/WRC (Cavity 1 – 633 nm).

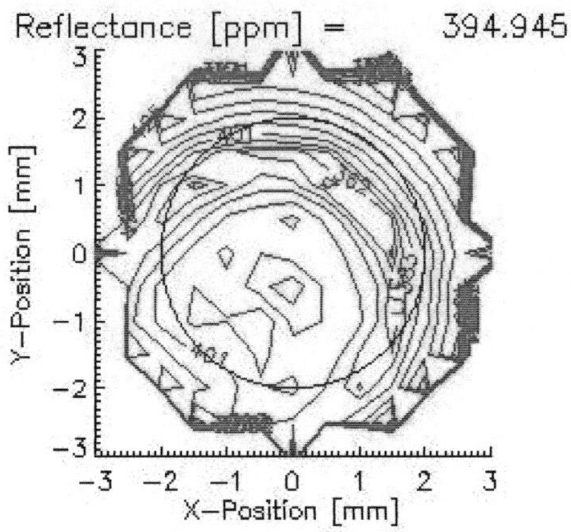

Figure 2. - Example of a measurement done by NIST (Cavity 1 – 660 nm).

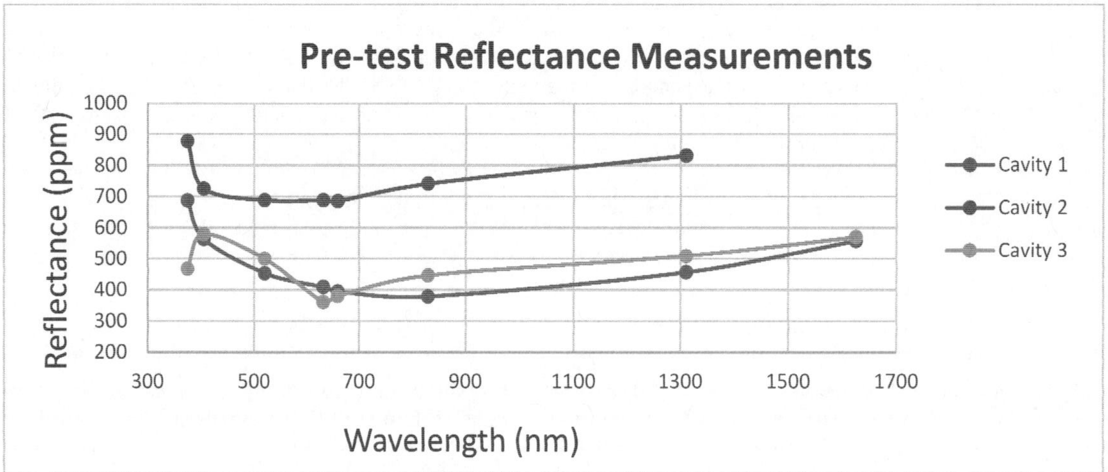

Figure 3. Summary of reflectance measurements before the tests.

The reflectance measurements performed at PMOD/WRC agree well with the measurements performed with a similar integrating sphere setup at LASP as can be seen in Figure 3.
These reflectance values were used as a reference to compare them with the reflectance measurements after the tests and will give us information about potential degradation occurring in the cavities because of the thermal cycling or the UV treatment.

The discrepancy in reflectance shown for Cavity 2 in contrast to 1 and 3 is attributable to Cavity 2 having only one layer of carbon nanotubes against the two layers in the remaining cavities. Three layer coatings were not tested, but a third layer would not improve the reflectance values according to the manufacturer.

IOP Conf. Series: Journal of Physics: Conf. Series **972** (2018) 012007 doi:10.1088/1742-6596/972/1/012007

3. Scanning Electron Microscope

In addition to the reflectance measurements, SEM imaging provides a qualitative study of the carbon nanotubes to identify any coating defects inside the cavities. Before and after the cavity treatment (either the ageing or the thermal cycling), SEM pictures of the surface of the coating were taken for a qualitative impression of how the mentioned treatments affect the surface of the sprayed paint and the structure of the carbon nanotubes.

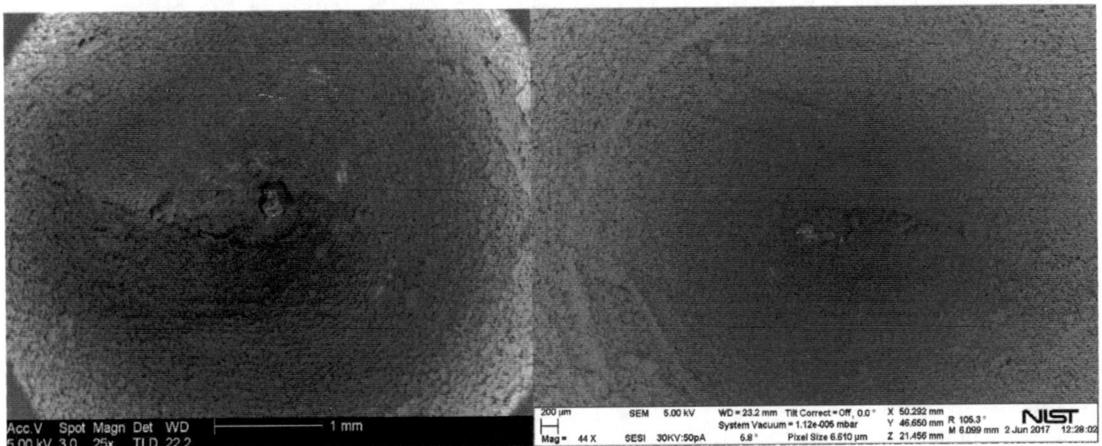

Figure 4. SEM pictures of Cavity 1 (left: pre-treatment; right: post-UV treatment). No significant differences are visually detected.

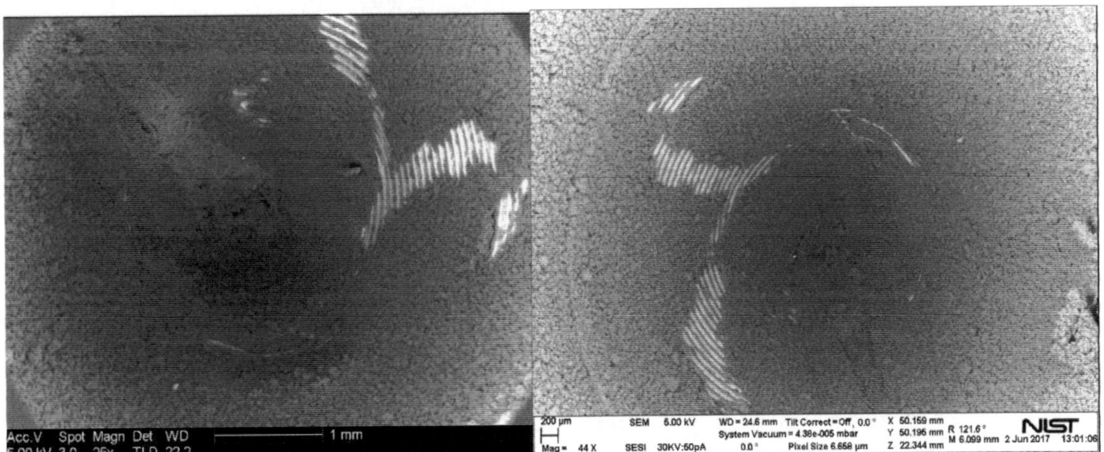

Figure 5. SEM pictures of Cavity 2 that remains as a reference (left: pre-treatment; right: post-treatment of the other cavities). Defects in the walls of the cavities attributable to a wrong handling of the detector, which results in a higher reflectivity.

Because of the binder used in the cavities, a commercial glue that fixed the heater foil inside the cavity before applying the carbon nanotubes coating, the process temperature for spraying the carbon nanotubes was limited, obtaining the meniscus as a result that can be seen in Figures 4 and 6. This does not represent a problem in the properties of the detector, but it is something to be optimized by looking for components with a higher temperature resistance. The currently used binder for the heater foil only allows temperatures up to 120°C, whereas the typical application temperature for the sprayable CNT's is 280°C.

IOP Conf. Series: Journal of Physics: Conf. Series **972** (2018) 012007 doi:10.1088/1742-6596/972/1/012007

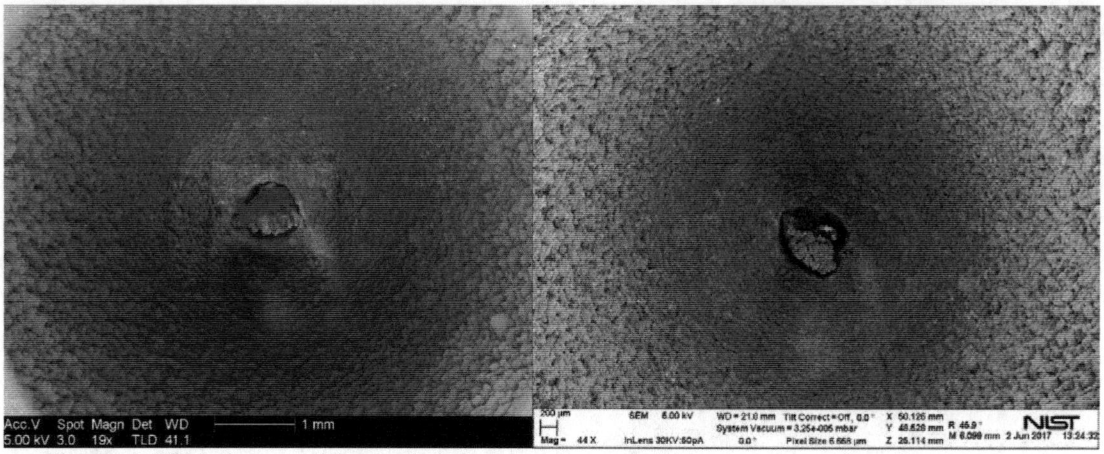

Figure 6. SEM pictures of Cavity 3 (left: pre-treatment; right: post- thermal cycling treatment).

4. Thermal cycling

Thermal cycling was performed for Cavity 3 to evaluate any gross deficiencies in the assembly (including both coating and substrate) preliminary to eventual space qualification testing. Our radiometer was subjected to thermal cycling for 21 cycles over 6 days. Each cycle was 6 hours and 30 minutes, with a maximum temperature of 80 °C and a minimum temperature of -20 °C.

The cycle was thermally controlled by a thermistor inside the chamber every minute to check the stability of the test.

This preliminary thermal cycling was not seen to induce coating cracks and with their respective reflectivity changes. The cavity structure demonstrated no degradation.

5. UV exposure

Cavity 1 was subjected to a high-power UV laser ($\lambda = 248$ nm) as our goal was to simulate the equivalent UV dose for one year of full time solar exposure. The setup, shown in Figure 7, has a 0.8 cm diameter aperture to limit the UV laser beam diameter and the sample was installed inside a vacuum chamber maintained below $6x10^{-4}$ mbar and the cavity was thermally controlled by a thermistor with a maximum temperature of 85 °C.

A beam splitter reflected 3.8% of the incident radiation to a monitor detector and allowed us to have a real-time measurement of the pulse.

The cavity radiometer was exposed during 5 hours to 3.6 million pulses of approximately 20 mJ/pulse with a frequency of 200 pulses/s, with an average irradiance of 2 W/cm^2.

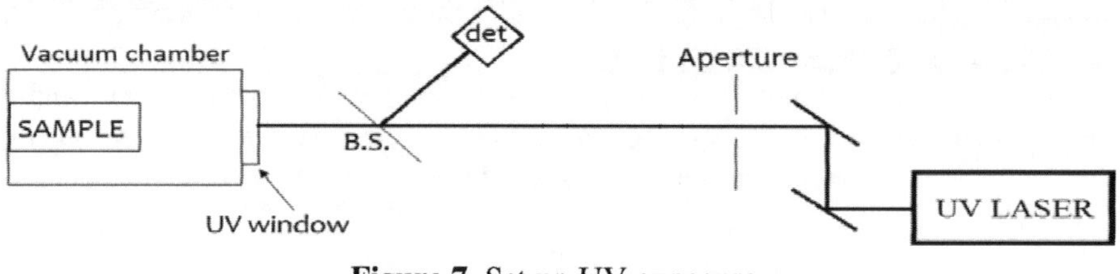

Figure 7. Set up UV exposure.

First, we crudely estimate the temperature rise that might show in the carbon nanotubes due to the pulsed laser by assuming 1) each pulse is absorbed by the nanotubes only without conduction to the substrate, 2) the absorbed energy propagates to the substrate before the arrival of the next pulse, and 3) the substrate temperature rise is trivial. The temperature rise of the nanotubes is estimated, using the familiar heat capacity formula:

$$E = m\, C\, \Delta T \tag{1}$$

Here *m* is the mass of the carbon nanotubes in the cavities with two layers; 5 mg according to the manufacturer, the specific heat capacity *C* can be compared for the value in [4] for MultiWall carbon nanotubes (MWCNT) and is 8 *J/mol K* and the energy of the laser *E* is 20*mJ*.
Using these values in (1) we find a temperature rise of 6 K for each laser pulse for the carbon nanotubes. This value suggests that the peak of temperatures during each pulse will increase the temperature, and, as it is indicated by the manufacturer, a temperature higher than 325°C would have broken the hydrophobic protection film, consequently damaging the coating and increasing the reflectivity.

6. Results and Discussion

After the different treatments, we again measured the reflectance of the cavities and we compared them to the measurements before the treatments in Figure 8.

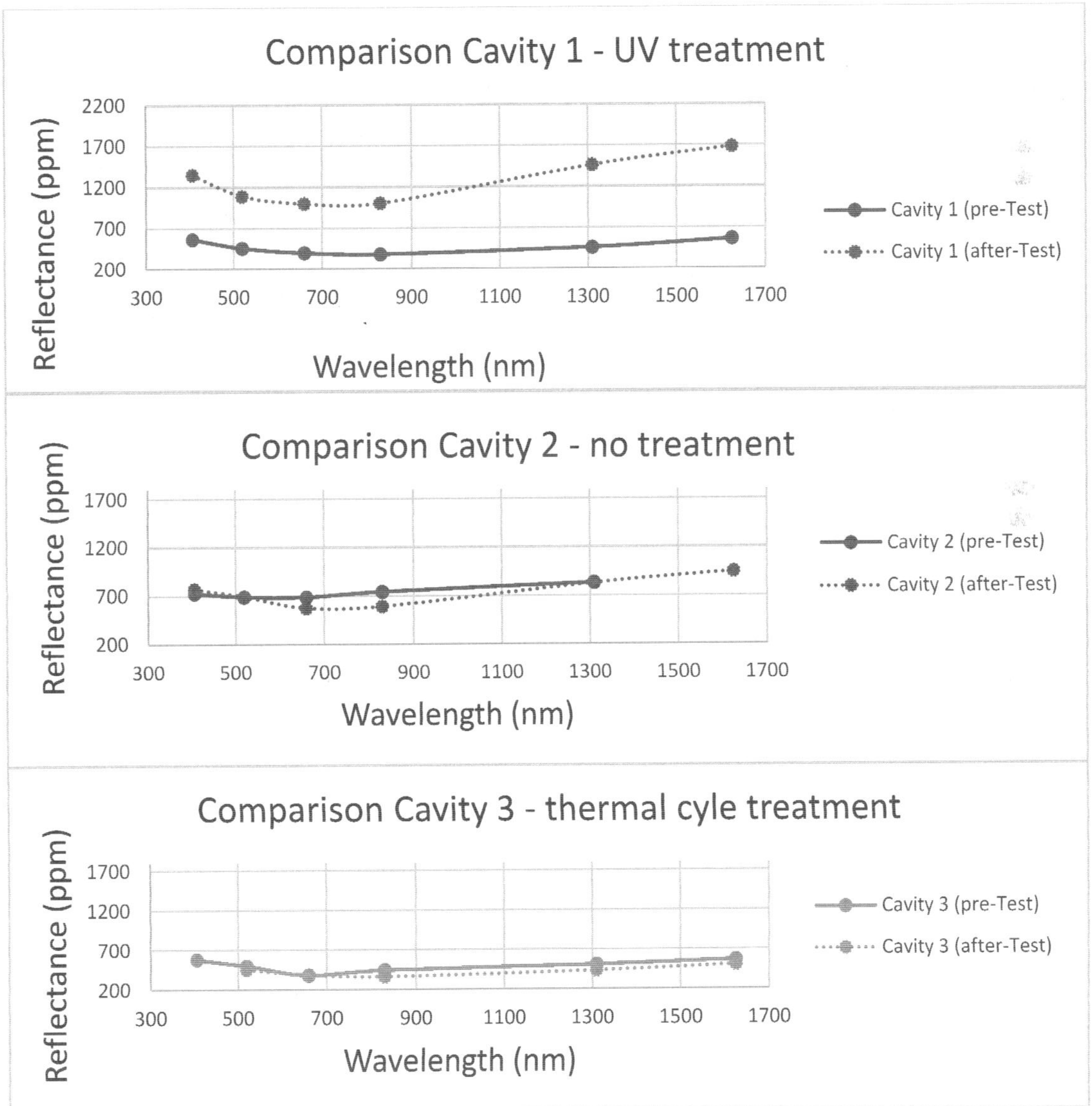

Figure 8. Comparison of the three cavities previously and after each treatment by LASP.

Cavity 2, which served as a reference, showed a decrease in the reflectance values even though when no treatment was applied. This discrepancy is most likely attributable to experimental errors during the measurements after the treatments, smaller than 10%, described in [3].
Measurements of Cavity 3 before and after thermal cycling show no significant difference and indicate the thermal cycling has not affected the carbon nanotubes within the uncertainties shown for Cavity 2.

Although there is no significant difference in the SEM imaging (see figures 4 to 6), after UV laser accelerated aging, Cavity 1 suffered a significant increase in the coating reflectance, as can be seen in Figure 8. The temperature reached in the carbon nanotubes is most likely higher than what we measured in the cavity and might be above the temperature limit of the hydrophobic film. This overheating of the coating might induce breaks in the protection film and damage on the carbon nanotubes. We saw this behaviour, extremely pronounced, for a VACNT sample that had a thermal grease under the substrate. This effect will be discussed in another paper on, currently under development.
Another explanation could be outgassing produced by the carbon nanotubes coating inside the vacuum chamber which might have contaminated the carbon nanotubes.
For space applications, measuring TSI, the nanotubes will not be subjected to high UV pulse peaks, as we performed in the experiment. Therefore, the temperature in the carbon nanotubes will not be higher than the binder limit.
As well, outgassing of the detector will be performed before launching, thus potential contamination will be included.
On the other hand, we proved that thermal changes may not affect the detectors.

We performed the first steps in the characterization of this novel carbon-nanotube coating in a conical cavity for TSI. Further optimization of the detectors, such as a binder with higher temperature resistance or a new cavity shape optimized for the characteristics of the carbon nanotubes, may be applied. As well, further experiments for a better knowledge of the outgassing or analysis of the behaviour under environment with atomic oxygen will give us more information about the application of this new detectors for TSI.

7. References

[1] https://www.surreynanosystems.com/
[2] Walter B, Levesque P-L, Kopp G, Andersen B, Beck I, Finsterle W, Gyo M, Heuerman K, Koller S, Mingard N, Remesal Oliva A, Pfiffner D, Soder R, Spescha M, Suter M, and Schmutz W 2017 The CLARA/NORSAT-1 solar absolute radiometer: instrument design, characterization and calibration *Metrologia*
[3] Sutter M 2015 Advances in Solar Radiometry, *PhD thesis, University of Zurich*
[4] Muratov V B, Vasil'ev O O, Kulikov L M, Garbuz V V, Nesterenko Yu V, and Duda T I 2012 Thermodynamic properties of multiwalled carbon nanotubes *Journal of Superhard Materials*

Acknowledgments

This work is supported by SNF grant 200021 162926.
Thanks to the Sources and Detectors Group of NIST in Boulder (Colorado), specially to Michelle Stephens, Brian Simonds, Chris Yung, David Livigni, and John Lehman for opening their laboratories to me, and for investing their time and effort for this work.

IOP Conf. Series: Journal of Physics: Conf. Series **972** (2018) 012008 doi:10.1088/1742-6596/972/1/012008

The absolutely characterized nitrogen vacancy center-based single-photon source – measurement uncertainty of photon flux and angular emission properties

B Rodiek, M López, H Hofer and S Kück

Physikalisch-Technische Bundesanstalt (PTB), Bundesallee 100, 38116 Braunschweig, Germany

beatrice.rodiek@ptb.de

Abstract. Recently, a single-photon source based on the nitrogen vacancy center in a nanodiamond with a traceable spectral photon flux was realized [1]. In this article, we report on the determination of the measurement uncertainty of the spectral photon flux (about 4 %) and discuss its different components. Furthermore, the angular distribution of the emission of an NV center in a nanodiamond located in the vicinity of a dielectric boundary is calculated based on the model of Lukosz and Kunz. The agreement between the angular distribution calculation and the first-ever measurement of the angle-dependent emission in nanodiamond NV centers is satisfactory.

1. Introduction

Single-photon sources (SPSs) are highly relevant for several forms of quantum-based technology such as quantum communication and quantum key distribution [2], [3], [4]. In radiometry, SPSs with a negligible background and a pure single-photon emission show promise as standard sources [1], [5] that can efficiently calibrate single-photon detectors; however, this is because the influence of the photon statistics on the calibration results is omitted [1], [6]. Currently, single-photon detectors are calibrated by means of attenuated lasers, resulting in an unavoidable measurement correction caused by the photon statistics and the dead time of the detector [7]. Using a pure, on-demand single-photon source would make such a measurement correction unnecessary [6], thus leading to lower uncertainties. A possible candidate for such a single-photon source is the nitrogen vacancy (NV) center in nanodiamond, which photon emission has shown to be stable at room temperature (i.e., it does not bleach or show photo-blinking) [4].

In this paper, we briefly summarize the metrological characterization of a single-photon emitter with respect to its total spectral photon flux, as presented in detail in Ref. [1]. Furthermore, we present the detailed estimation of its measurement uncertainty and report on the calculation and measurement of the angular emission characteristic of an NV center.

2. The single-photon source

The single-photon source consists of the emitter and the setup that excites the emitter and collects and analyzes its emission. A nanodiamond doped with a nitrogen vacancy center (NV center) was used as the emitter. The nanodiamond had a size of approximately 75 nm and was placed on a cover glass by means of a spin coating procedure. The sample was fabricated by Friedrich Alexander University (FAU).

The setup used is a confocal microscope developed in-house [1]. Excitation takes place via a laser beam at a wavelength of 532 nm, which is reflected by a dichroic beam splitter and focused by means of an oil-immersion microscope objective (numerical aperture of 1.45). The light emitted passes the microscope objective and the dichroic beam splitter as well as some edge filters in order to ensure that only the NV center emission is collected, and in order to prevent any stray light from being transmitted. The emission by the NV centers is coupled into an optical fiber that is connected to several devices (i.e., a calibrated low optical flux detector (LOFD), a spectroradiometer and a Hanbury-Brown and Twiss interferometer (HBTI)) in order to determine the photon flux and the spectral distribution, and in order to check the single-photon emission, respectively [1]. All measurements are carried out at room temperature.

3. Absolute metrological characterization

The absolute metrological characterization of the NV center-based single-photon source was described in detail in [1]. The absolute photon flux per wavelength and per second is described by the following equation [1]:

$$N_{ph,\lambda}(\lambda) = \frac{\Phi_\lambda(\lambda)}{e_{ph}(\lambda)} = \frac{\lambda}{hc}\frac{U_{out}C_{int}}{t_{int}}\frac{S_{rel}(\lambda)}{\int S_{rel}(\lambda)s_{det}(\lambda)d\lambda} = \frac{\lambda}{hc}\frac{U_{out}C_{int}}{t_{int}s_{spec}}S_{rel}(\lambda)\,, \qquad (1)$$

where the factors h and c are the Planck constant and the speed of light, respectively. The measurands of the low-optical flux detector (LOFD) [8] are the output voltage U_{out}, the integration capacity C_{int} and the integration time t_{int}. The effective spectral responsivity s_{spec} is defined as the combination of the relative spectral distribution of the NV center emission $S_{\text{rel}}(\lambda)$ and the spectral responsivity of the detector $s_{\text{det}}(\lambda)$. $S_{\text{rel}}(\lambda)$ was determined by calibrating the spectrometer with respect to its spectral irradiance responsivity (i.e., traceable to the blackbody radiator) [1]. For clarification, the traceability chain for the determination of the spectral photon flux is shown in Figure 1a. The absolute spectral photon flux (i.e., the absolute number of photons per wavelength and per second) is shown in Figure 1b.

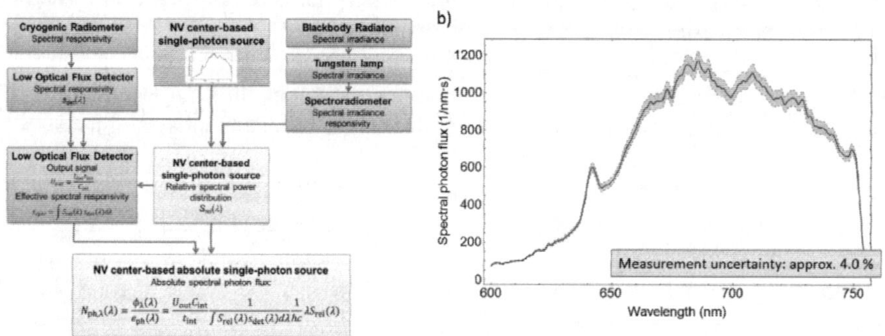

Figure 1: a) Traceability chain for the metrological characterization of a single-photon source in terms of its absolute spectral photon flux [1]; b) Absolute spectral photon flux of the single-photon source (blue curve). The calculation of the measurement uncertainty presented (red stripe) is described in Section 4.

4. Measurement uncertainties

The determination of the measurement uncertainty for the absolute spectral photon flux $N_{\text{ph},\lambda}(\lambda)$ follows the model shown in equation (1) and is carried out in accordance with the procedure described in the *Guide to the expression of uncertainty in measurement* (GUM) [9].

Several parameters have to be taken into account in order to determine the measurement uncertainties of the absolute spectral photon flux of the single-photon source. Three groups of parameters can be identified: uncertainties related to the error contribution from the determination of the spectrum; those related to the detection of the absolute photon flux; and some other, minor contributions.

The error caused by the spectrum is the largest. It mainly arises from the resolution error, which depends on the entrance slit and the grating used. These lead to smoothed structures that have to be corrected. Based on a measurement obtained by means of a narrow-linewidth laser, a triangular slit function with a bandwidth of 1 nm is assumed. In principle, this leads to a correction to the measured spectrum of about 3.8 %; for simplification, this is taken into account as a measurement uncertainty. Further analysis requires a more detailed characterization of the spectroradiometer, especially of the slit function that has to be carried out in the future. Furthermore, the effect of the spectrometer calibration has to be considered. Calibration with a standard tungsten lamp leads to an uncertainty in the relative spectral responsivity of about 1 %. Stray light and second-order effects are negligible: in the measurement with a narrow bandwidth laser, neither of these effects had a significant influence. This gives a total measurement uncertainty for S_{rel} of 3.9 %.

In addition to the spectral contributions, the detector also has an impact on the measurement uncertainty. The uncertainty sources associated with the LOFD are obtained from Ref. [8] and are thus mentioned only briefly below. The standard uncertainty of the effective spectral responsivity is 0.3 %. The integration time has a standard uncertainty of 0.3 %. The uncertainties related to the outcome voltage and the capacity are 0.5 % and 0.014 %, respectively [3]. All uncertainty sources and their contributions to the measurement uncertainty of the source photon flux for a wavelength of 770 nm are listed in Table 1.

Table 1: Contributions to the measurement uncertainty of the absolute spectral photon flux; U_{out} corresponds to the output voltage, C_{int} to the capacity, S_{rel} to the relative spectral power distribution, t_{int} to the integration time, and s_{spec} to the spectral responsivity.

Measurand	Value	Standard uncertainty (%)	Distribution	Contribution (%)
U_{out}	0.01854642	0.5	Standard	9.97
C_{int}	1.0046E-12	0.014	Standard	0.27
S_{rel}	0.0047	3.9	Square	77.79
t_{int}	1	0.3	Standard	5.98
s_{spec}	0.338088662	0.3	Standard	5.98
$N_{ph,\lambda}(\lambda)$		3.95	Standard	

As expected, the largest contribution of 77.8 % is caused by the measurement of the relative spectral power distribution S_{rel}. Overall, the relative standard uncertainty was calculated as being 4.0 % ($k = 1$), giving an expanded uncertainty ($k = 2$) of 7.9 %.

5. Angular emission of the NV center emission

The emission of the NV center in a nanodiamond can be considered a dipole emission into the 4π solid angle. However, it can be modified by placing the NV emitter in dielectric structures that alter the emission characteristic – for example, within the near-field distance of an antenna that could then improve the collection efficiency [10], [11].

The angular emission of the NV center in a nanodiamond is influenced by the cover glass, which is a dielectric structure and therefore acts like an antenna. This implies that the NV center emission is re-directed in a preferred direction. The calculation regarding the angular emission is carried out in accordance with reference [12]. The field of illumination is divided into three areas (see Figure 2a) and the results regarding the illumination intensity are rotationally symmetric around the z-axis. The first partition belongs to the transition between the air and the cover glass. Moreover, the illumination in the cover glass is divided into two areas that are separated by the critical angle (i.e., the angle of total reflection). Below, the angular emission calculations for the three areas in accordance with Ref. [12] are presented.

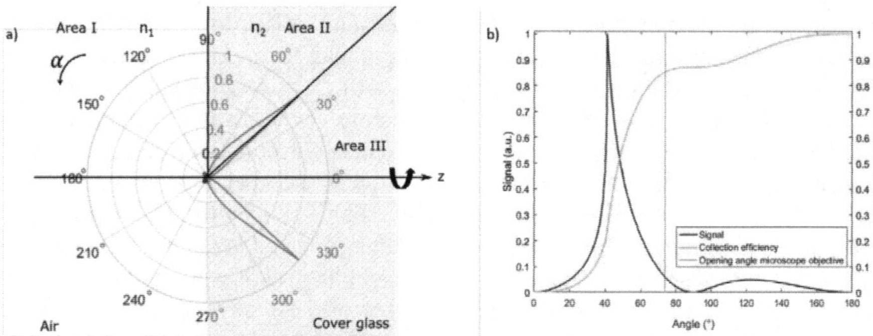

Figure 2: a) Schematic representation of the field of illumination divided into the three calculation areas; the calculated angular emission is presented in blue; b) Calculated angular emission of an NV center-doped nanodiamond on a cover glass (blue curve); the collection efficiency is presented in green and the opening angle of the microscope objective used is indicated in orange.

The electric dipole component of the emission into air (Area I in Figure 2a), $P_{e,L}$, is given by the following equation [12]:

$$P_{e,L} = \frac{3}{8\pi} \frac{n^2 sin^2(2\alpha_{L1})}{(n\cos(\alpha_{L1}) + cos(\alpha_{L2}))^2}. \quad (2)$$

The factor n is the ratio of the refractive index of the cover glass ($n_2 = 1.5255$) and the refractive index of air ($n_1 = 1$). The angles α_{L1} and α_{L2} are defined as follows [12]:

$$\frac{\pi}{2} \leq \alpha_{L1} \leq \pi \quad (3)$$

and

$$\alpha_{L2} = \arcsin\left(\frac{sin(\pi - \alpha_{L1})}{n}\right). \quad (4)$$

The second partition occurs in the cover slip itself at the angle of total reflection. For angles larger than the critical angle (Area II in Figure 2a), the equation for the electric dipole emission is [12]:

$$P_{e,G,2} = \frac{3}{8\pi} \frac{n^5}{n^2-1} \frac{sin^2(\alpha_{2G})}{((n^2+1)^2 sin^2(\alpha_{2G})-1)} \quad (5)$$

with

$$\arcsin\left(\frac{1}{n}\right) \leq \alpha_{2G} \leq \frac{\pi}{2}. \quad (6)$$

For angles smaller than the critical angle (Area III), the electric dipole emission is defined as [12]:

$$P_{e,G,1} = \frac{3}{8\pi} \frac{n^5 sin^2(2\alpha_{1G1})}{(n\cos(\alpha_{1G2}) + cos(\alpha_{1G1}))^2}, \quad (7)$$

where

$$0 \leq \alpha_{1G1} \leq \arcsin\left(\frac{1}{n}\right) \quad (8)$$

and

$$\alpha_{1G2} = \arcsin(n sin(\alpha_{1G1})) \, . \tag{9}$$

The results of the calculations are presented in Figure 2b (blue curve). The highest amount of the emission is directed into the cover glass at an angle of 41 °. The calculated collection efficiency (i.e., the integration over the signal normalized to the total signal) is also illustrated as a green curve. Here, 87 % of the illumination is emitted into the cover glass (angles from 0° to 90°) and only approx. 13 % (angles from 90 ° to 180 °) of the photons are directed into the air. The microscope objective used has a numerical aperture corresponding to a half opening angle of 72.8 ° (orange line), thus allowing 85 % of the emitted photons to be collected.

In the experiment, the angular emission is measured under the application of a back focal plane lens and an sCMOS camera. The resulting image is shown in Figure 3a. The pattern presented corresponds to two times the opening angle of the microscope objective. The intensity of the emission is color coded (i.e., the bright ring belongs to the strongest emission). The emission intensity profile was determined by taking the signal of the camera along the white line. As can be seen, the resulting plot (green curve in Figure 3b) agrees well with the calculations (blue curve). The deviation observed between the calculated and measured intensity profile is assumed to be due to distortion arising from the back focal plane lens.

Furthermore, the emission shows a circular intensity modulation, as can be clearly seen in the bright circle. This pattern indicates the dipole characteristic of the NV center emission and is observed for all NV centers under investigation. However, the orientation differs from one NV center to another, according to the orientation of the nanodiamond and the NV center.

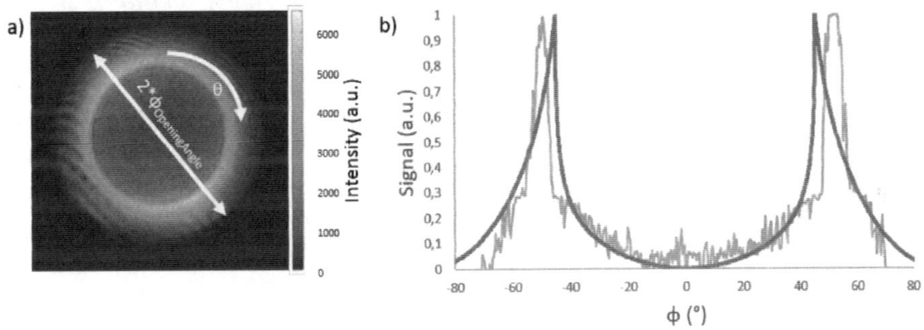

Figure 3: a) Measured picture of the angular emission of an NV center; b) Measured (green) and calculated (blue) intensities of the angular emission for the opening angle of the microscope objective in the range of -80° to 80°.

6. Conclusion

In this article, the metrological characterization of a single-photon source based on a nitrogen vacancy center-doped nanodiamond was presented in terms of its absolute spectral photon flux. The relative standard uncertainty of the spectral photon flux was determined to be 4.0 %. To reduce the uncertainty, it is planned to optimize the resolution of the spectroradiometer in order to lower the highest uncertainty contribution.

Furthermore, investigations on the angular emission of an NV center were presented. The results of the measurement correspond to the calculations and the dipole characteristic of the NV emission was observed. The measurement carried out was for a half sphere only; for this reason, further investigations are planned in which the angular emission in a full sphere will be measured by means of a double-confocal microscope [13].

References

[1] B. Rodiek, M. Lopez, H. Hofer, G. Porrovecchio, M. Smid, X.-L. Chu, S. Gotzinger, V.

Sandoghdar, S. Lindner, C. Becher, and S. Kuck, "Experimental realization of an absolute single-photon source based on a single nitrogen vacancy center in a nanodiamond," *Optica*, vol. 4, no. 1, p. 71, 2017.

[2] N. Sangouard and H. Zbinden, "What are single photons good for?," *J. Mod. Opt.*, vol. 59, no. September, pp. 1–7, 2012.

[3] G. S. Buller and R. J. Collins, "Single-Photon Generation and Detection," *Meas. Sci. Technol.*, vol. 21, no. 1, p. 28, 2010.

[4] M. D. Eisaman, J. Fan, J. Migdall, and S. V Polyakov, "Invited Review Article: Single-photon sources and detectors," *Rev. Sci. Instrum.*, vol. 82, no. 71101, 2011.

[5] W. Schmunk, M. Gramegna, G. Brida, I. P. Degiovanni, M. Genovese, H. Hofer, S. Kück, L. Lolli, M. G. A. Paris, S. Peters, M. Rajteri, A. M. Racu, A. Ruschhaupt, E. Taralli, and P. Traina, "Photon number statistics of NV centre emission," *Metrologia*, vol. 49, no. APRIL, pp. S156–S160, 2012.

[6] C. J. Chunnilall, I. Pietro Degiovanni, S. Kück, I. Müller, A. G. Sinclair, C. J. C. Ivo, P. Degiovanni, S. Kück, I. Müller, A. G. Sinclair, C. J. Chunnilall, and I. Pietro Degiovanni, "Metrology of single-photon sources and detectors: a review," *Opt. Eng.*, vol. 53, no. 8, p. 81910, 2014.

[7] M. López, H. Hofer, and S. Kück, "Detection efficiency calibration of single-photon silicon avalanche photodiodes traceable using double attenuator technique," *J. Mod. Opt.*, vol. 62, no. 20, pp. 1732–1738, 2015.

[8] G. Porrovecchio, M. Šmid, M. López, H. Hofer, B. Rodiek, and S. Kück, "Comparison at the sub-100 fW optical power level of calibrating a single-photon detector using a high-sensitive , low- noise silicon photodiode and the double attenuator technique," *Metrologia*, vol. 53, pp. 1115–1122, 2016.

[9] J. C. F. G. I. Metrology, *Evaluation of measurement data — Guide to the expression of uncertainty in measurement*. 2008.

[10] X.-L. Chu, T. J. K. Brenner, X.-W. Chen, Y. Ghosh, J. A. Hollingsworth, V. Sandoghdar, and S. Götzinger, "Experimental realization of an optical antenna designed for collecting 99% of photons from a quantum emitter," *Optica*, vol. 1, no. 4, pp. 203–208, 2014.

[11] D. Zhang, R. Badugu, Y. Chen, S. Yu, P. Yao, P. Wang, H. Ming, and J. R. Lakowicz, "Back focal plane imaging of directional emission from dye molecules coupled to one-dimensional photonic crystals," *Nanotechnology*, vol. 25, no. 14, 2014.

[12] W. Lukosz and R. E. Kunz, "Light emission by magnetic and electric dipoles close to a plane dielectric interface. II. Radiation patterns of perpendicular oriented dipoles," *J. Opt. Soc. Am.*, vol. 67, no. 12, pp. 1615–1619, 1977.

[13] B. Rodiek, M. López, H. Hofer, K. Dhoska, and S. Kück, "Double Scanning Confocal Microscope for the characterization of single- photon emitters," in *Single Photon Workshop 2015, www.unige.ch/spw2015/_media/submissions:abstracts:spw2015_submission_109.pdf*, 2015.

Acknowledgements

This work was funded by the project "Single-Photon Sources for Quantum Technology" (SIQUTE) of the European Metrology Research Programme (EMRP). The EMRP is jointly funded by the EMRP participating countries within EURAMET and the European Union.

This work has been supported by EMPIR-14IND05 "MIQC2" (the EMPIR initiative is co-funded by the EUH2020 and the EMPIR Participating States).

We gratefully acknowledge support by the Braunschweig International Graduate School of Metrology B-IGSM and the DFG Research Training Group GrK1952/ 1 "Metrology for Complex Nanosystems" and Peter Sperfeld from the "Spectroradiometry" working group of PTB.

IOP Conf. Series: Journal of Physics: Conf. Series **972** (2018) 012009 doi:10.1088/1742-6596/972/1/012009

High power LED standard light sources for photometric applications

Evgeniy Ivashin, Sergey Ogarev, Boris Khlevnoy, Stanislav Shirokov, Dmitry Dobroserdov and Victor Sapritsky

All Russian Research Institute for Optical and Physical Measurements (VNIIOFI), Moscow, Russia

Corresponding e-mail address: ivashin@vniiofi.ru

Abstract. High power LED light sources have been developed as possible new VNIIOFI standard sources for luminous intensity, luminous flux and colour measurements. Stability, repeatability and spatial uniformity of the sources were investigated and demonstrated high accuracy and homogeneity. The paper describes different tests on one of the manufactured sources. In the future, these LED light sources are planned to be used as standard luminous flux sources to transfer the units of luminous intensity and luminous flux from gonio-spectrometer to sphere-spectrometer.

1. Introduction

For the purpose of dissemination of the units of luminous intensity and luminous flux, incandescent lamps are currently used. The use of incandescent lamps involves a number of difficulties: transportation, large heat emission during operation (for large levels of luminous intensity and luminous flux), the effect of ambient temperature on metrological characteristics, and relatively short service life. The main trend at present is the use of light-effect-diodes (LEDs) for transfer of a unit of luminous intensity and luminous flux. Growing-up LED lighting industry and lack of incandescent lamps rises the need for developing of new photometric standards on the basis of LEDs. Activities aimed at such development are carried out at CCPR, CIE and a number of NMIs [1-4]. In particular, in the presentation [1] of Dr. Y. Ohno at the 23rd Meeting of CCPR (Sevres, France) in September 2016, dedicated to LED sources in Photometry at NIST, the following related activities in CCPR are mentioned: (1) Working group of KC TG-4 Pilot study for the use of alternative standards for photometric comparisons, which is devoted to investigate the use of white LED products (but not limited to those) as artifacts for future photometric comparisons in CCPR; (2) Working group of SP Discussion Forum on use of white LED sources for photometry.

VNIIOFI has also started a project on developing and investigating high power LED light sources, which are to be used as standard sources for luminous intensity, luminous intensity distribution, luminous flux and colour measurements. To transfer a unit of luminous intensity and luminous flux, LED sources on the basis of powerful chip on board (COB) diodes and a control device for these sources were developed to replace SIS 107-500, SIS 107-1000, and SIP 107-1500 lamps.

The design of the source, its stability investigation results, spectral spatial uniformity, and luminous intensity spatial uniformity are presented in this paper. Further, in this work we describe in few words VNIIOFI realization of power supply for precision stabilizing of high power LED sources.

2. Design of LED source

The source consists of two parts: a light head and a parametric current source. The design of the light head (Figure 1) allows using any LEDs of the COB type. Three heads were built: two white with power of 36 W and 68 W and a blue one of 30 W. White LEDs were selected as the most stable from a batch after seasoning for 250 hours. When the sources were assembled they were additionally seasoned for 100 hours. The blue LED was manufactured specially for VNIIOFI and was pre-aged by manufacture. For reference sources, LEDs were selected for the following parameters: spatial distribution of luminous intensity, correlated color temperature, luminous intensity, luminous flux, operating temperature range at rated current, type of substrate.

The light head contains a Peltier element and a temperature sensor Pt1000 for high-accuracy temperature stabilization of the LED. The choice of platinum thermoresistance as a diode temperature control sensor is due to the high stability and the possibility of using this type of sensors in the bridge circuit, which is necessary to control the diode temperature in the selected stabilization control circuit. Peltier element was selected with double power reserve on cooling productivity. The LED is surrounded by a diffuse wall chamber consisting of a PTFE tube and a transparent opal glass, which makes angle radiation homogeneous with a good approach to the lambert law and makes spectral distribution independent from the angle of observation. Moreover , using the PTFE tube and opal glass establishes a volume like a «temperature chamber» surrounding the LED, which is a good location for the temperature sensor. LEDs have a special point with minimal thermal resistance to light emitting crystals on the same side with them. In front of the opal glass there is an aperture with diameter of 20 mm. All connections of elements placed on different sides of the Peltier element are made with special thermo isolated screws. Heatsink with cooler was selected by power dissipation according to Peltier electrical power. All orange elements in fig.1 are made from copper, because it has well enough thermal conductivity.

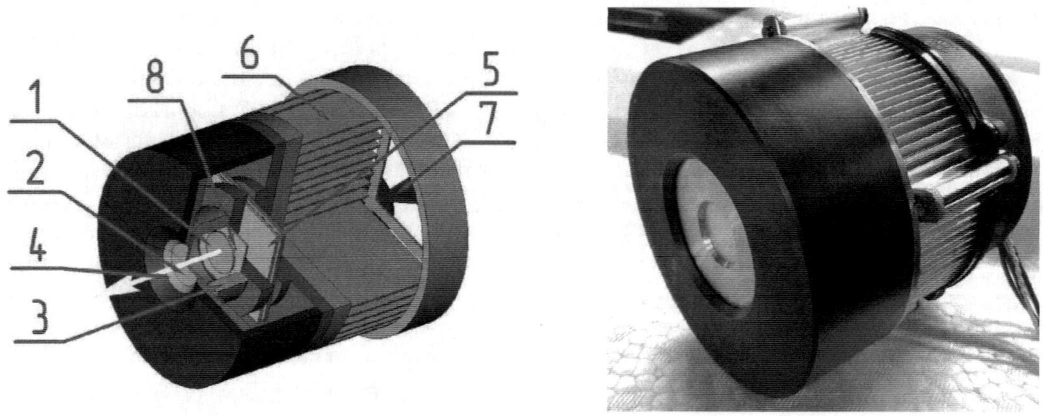

Figure 1. Design of light-head. 1. High power chip on board (COB) LED, 2.Opal glass, 3. PTFE tube, 4. Aperture, 5. Peltier, 6. Heatsink, 7. Fan, 8. Pt1000.

The values of axis luminous intensity of the built white LED sources are approximately 650 cd and 1000 cd, and the values of luminous flux are 1500 lm and 2500 lm.

The second part of the stabilized LED light sources is fine enough power supply for the LED, and a power supply with temperature measurements from Pt1000 and feedback to regulate the current of the Peltier depending on the temperature. Each power supply in VNIIOFI standard LED supply is a stabilized current sources. Both the power supply and the temperature measurement system are realized in one package and named also parametric power supply.

The block diagram of the parametric power supply is shown in Figure 2. It consists of two independent current sources (one for LED and another for Peltier), a temperature measurement system, a control block for realizing feedback for temperature stabilization and one fixed source with 12 volts up to 0.5 ampere for cooler on heatsink in LED construction (Figure 1). The current source supplying the LED is a stabilized source with a discreteness of the current setting of $30 \cdot 10^{-6}$A. The discreteness of the setting of the current generated by the power source of the Peltier element is $50 \cdot 10^{-6}$ A. The small step of setting the Peltier current is due to the need for high-precision stabilization of the diode temperature with a small change in the ambient temperature. The control unit generates a control signal for the power source of the Peltier element based on the diode temperature. Besides, power supply system has connection port RS232 to control the system via computer. Each current channel provides current up to 2.1 A with resolution of 0.03 mA. The range of the stabilized temperatures can vary from 30°C to 45 °C with resolution of 0.01 °C. For the LEDs mentioned above, the temperature of 36 °C was used.

Typical current for the chosen LED`s is about 1 ampere. Stability of the current source for LED on 1 ampere level is the same as its resolution, 0.03 mA.

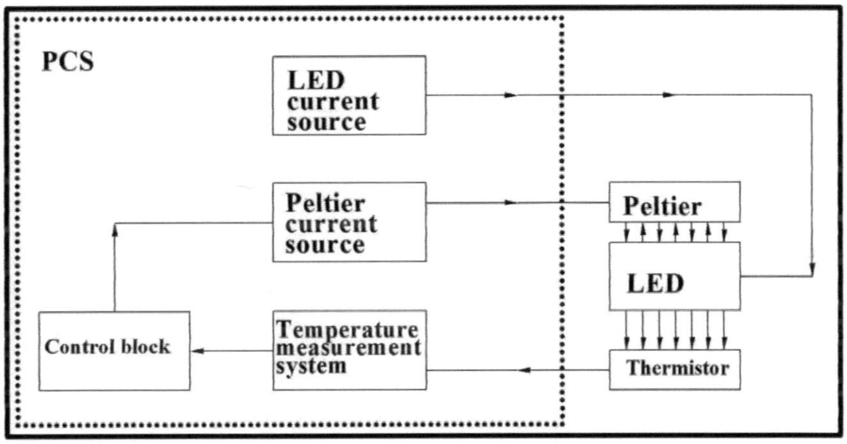

Figure 2. Flowchart of power supply system (parametric current sources).

3. Investigation of stability

At the first stage of investigation of the new sources their stability and repeatability were determined. First the temperature stability of the diode was checked for different levels of ambient temperature. The result of this test is shown in Figure 3. The LED temperature was stable within ± 0.05°C for 3.5 h while the ambient temperatures gradually varied from 20°C to 33°C and ±0.2°C for the range from 33°C to 36°C. This experiment was carried out to make sure that the cooling productivity of the chosen Peltier will be enough to stabilize the temperature of the LED in quite wide temperature range. Wide ambient temperature range is important because standard LED`s are planned to be used in industrial laboratories to calibrate their facilities.

Photometric stability of the source was investigated using a temperature stabilized LMT photometer equipped with a precision amplifier. The LED source and the photometer were placed in a light-tight room with ambient temperature of 22°C. The distance from the source to the photometer was 3.162 m. The luminous intensity of the LED was recorded automatically during a whole working day (Figure 4). The luminous intensity demonstrated slight drift of about 0.02% for 7 h. The reason of the observed drift could be stability of the LED power supply as well as the stability of the measurement system.

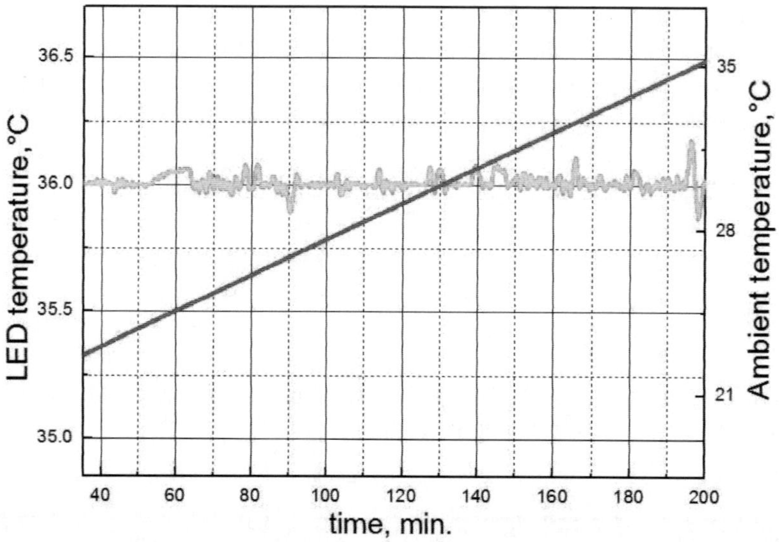

Figure 3. Relation of LED temperature stability from ambient temperature (green - LED temperature, red-ambient temperature)

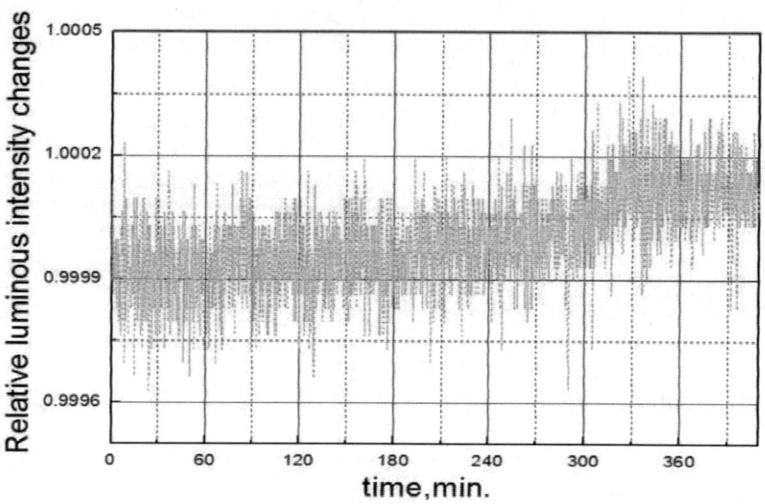

Figure 4. Relative change of luminous intensity.

Repeatability of the source was estimated by 10 measurements; each measurement was performed after turning the LED off, re-alignment and on again. Standard deviation of the measurements was 0.035% with drift of luminous intensity included.

4. Investigation of spatial homogeneity

The LED-based thermostabilized sources were investigated for their spatial homogenities of the luminous intensity distributions and spectral distributions in different angles with a C-γ goniometer including a thermo stabilized LMT photometer with low-current precision measurement system and a

IOP Conf. Series: Journal of Physics: Conf. Series **972** (2018) 012009 doi:10.1088/1742-6596/972/1/012009

spectroradiometer Konica Minolta CS-2000A. Spectral measurements were provided in reflectance mode using a small barium-coated plate.

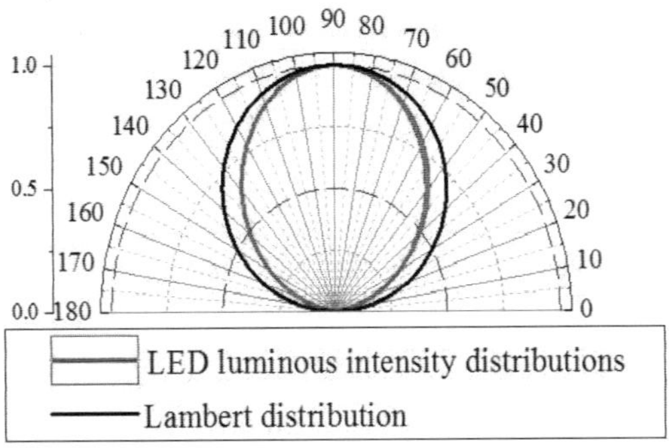

Figure 5a. Luminous intensity distribution with lambert distribution.

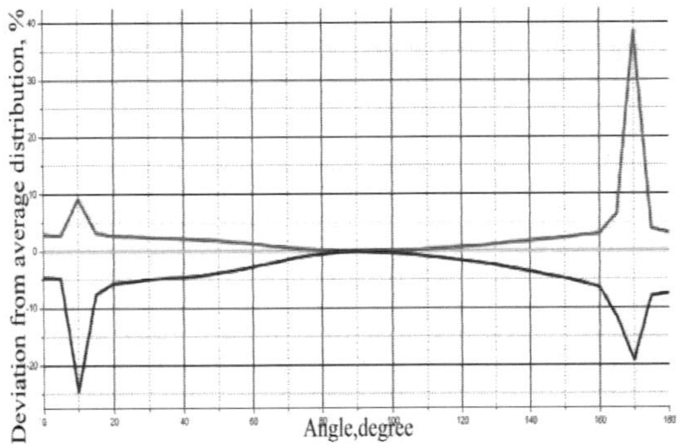

Figure 5b. Deviation from average distribution.

Spatial luminous intensity distribution of the source is shown in Figure 5a. Luminous intensity distribution has good homogeneity. Standard deviation of luminous intensity between different C-angles (for fixed γ-angles) is less than 2% for the range of γ-angles from -70° to 70°. This fact is very important because it made possible calibration of industrial gonio-spectrometers and gonio-photometers with bigger angular steps to decrease time of calibration. But tests show that form of aperture in front of the light-head is bad, because we saw reflectance from the aperture bevel (10° and 170° peaks on the graph in Figure 5b).

After spectral distributions measured in different directions colour coordinates were calculated and the results are shown in Figure 6.

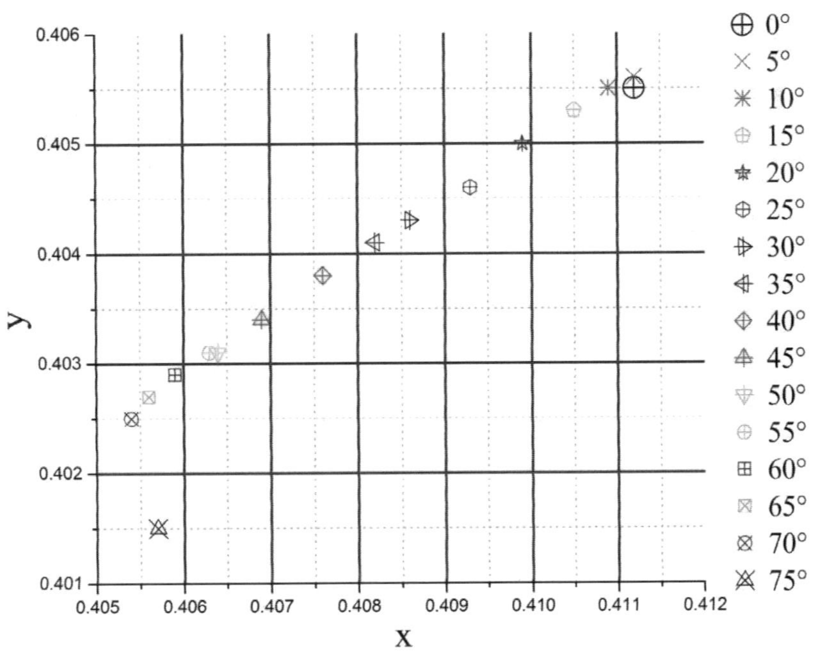

Figure 6. Relation between colour coordinates and angle of observation.

The maximum deviation of the colour coordinates is Δx=0.006 and Δy=0.004 from the 0° axis colour coordinates. This small deviation is simplifying calculations of uncertainties for the spectral mismatch correction factor when we calibrate gonio-photometers with this thermo-stabilized LED.

5. Conclusions

The design of quasi-lambertian high power LED sources for chip on board type LEDs has been developed. Two white LED sources have been built with luminous intensity of 500-1000 cd and luminous flux of 1500-2500 lm and photometric stability of 0.02% for approximately 10 h.

The developed LED sources can be used as standard photometric sources and may, in future, replace incandescent lamps.

Acknowledgments

The work was carried out using the equipment of the Center for the collective use of high-precision measuring technologies in the field of photonics (ckp.vniiofi.ru), created on the basis of FSUE VNIIOFI and supported by the Ministry of Education and Science of Russia in the framework of the implementation of agreement No. 14.595.21.0003 of August 28, 2017 (the unique identifier RFMEFI59517X0003).

References

[1] CCPR WG-KC TG-4 "Pilot study for the use of alternative standards for photometric comparisons"
[2] P. Blattner, CIE report to CCPR 2016
[3] Y. Ohno et al, LED Sources in Photometry at NIST, report to CCPR 2016
[4] T. Zama, NMIJ Research Activities relevant to LED Sources for Photometry, report to CCPR 2016
[5] T. Gerloff, M. Lindemann, S. Shirokov, M. Taddeo, S. Pendsa, A. Sperling; Development of a new high-power LED transfer standard; CIE Proceedings 2012.
[6] M. Lindemann, R. Maass; Photometry and Colorimetry of Reference LEDs by Using a Compact Goniophotometer; MAPAN – Journal of Metrology Society of India; Vol. 24; No. 3.
[7] CIE 127; 2007 (2nd edition); Measurement of LEDs.

LED based reference for wavelength and relative intensity

Hans Baumgartner[1], Kasperi Kylmänen[1], Olli Kantamaa[1], Petri Kärhä[1], and Erkki Ikonen[1,2]

[1] Metrology Research Institute, Aalto University, Espoo, Finland
[2] MIKES Metrology, VTT Technical Research Centre of Finland Ltd, Espoo, Finland

hans.baumgartner@aalto.fi

Abstract. Relative emission spectra of LEDs depend on the junction temperature. The high-energy region of the emission spectrum can be modelled with the joint density of states and the Maxwell-Boltzmann distribution as a function of energy and junction temperature. It can be shown that the normalized emission spectra at different junction temperatures intersect at a unique energy value. Thus the wavelength and the relative intensity of the intersection point do not depend on the junction temperature of the LED. The invariant intersection energy exists for all LEDs manufactured using the elements from groups III-V. The wavelength determined by the intersection energy can be used as a temperature invariant wavelength and relative intensity reference in spectral measurements.

1. Introduction

The emission spectrum of an LED is generated by electron-hole recombination and can be modelled as a product of the joint density of states of electrons and holes and the Maxwell-Boltzmann distribution [1,2]. The intensity and the peak wavelength of the light emitted by an LED depend on the junction temperature of the LED. An LED spectrum shows a redshift in the peak energy with increased junction temperature due to the reduced band gap with an increasing temperature [3]. When using LEDs in metrological applications as a wavelength reference, the junction temperature of the device needs to be stabilized in a reproducible way for different ambient temperatures.

It has been shown that the relative LED spectra, normalized to the peak intensity, have an energy value E_B, where the relative intensity is constant, i.e., independent on the junction temperature [1]. This value can be utilized as temperature-invariant wavelength and relative intensity characteristics for the measured LED. We have designed and built a device utilizing the temperature independent energy value as a wavelength and relative intensity reference.

2. LED spectrum model

The emission spectrum of an LED has been mathematically modelled earlier by numerous researchers [2,4,5]. The emission intensity of an LED is proportional to the product of the joint density of states and the exponential Boltzmann distribution of carriers. In [1], the temperature dependent LED spectrum was modelled as

$$I(E,T) \propto \left(E-\left(E_B - pkT\right)\right)^{p-q} \times e^{-\frac{E-(E_B-pkT)}{kT}}, \qquad (1)$$

IOP Conf. Series: Journal of Physics: Conf. Series **972** (2018) 012010 doi:10.1088/1742-6596/972/1/012010

where I is the emission intensity, E is the photon energy, T is the temperature of the LED p-n junction, E_B is the temperature-invariant intersection energy, p is a parameter defining the band gap energy shift as a function of temperature, q is a parameter defining the shift of the peak energy as a function of temperature, and k is the Boltzmann constant.

Equation (1) can be normalized by dividing with the intensity at the peak energy

$$E_{\max}(T) = E_B - qkT \,. \qquad (2)$$

From Eqs. (1) and (2), we can conclude that at the photon energy E_B the relative intensity does not depend on temperature [1]

$$\frac{I(E_B,T)}{I(E_{\max},T)} = \left(\frac{p}{p-q}\right)^{p-q} e^{-q} \,. \qquad (3)$$

The temperature dependence of the emission intensity of LEDs can be measured with a spectroradiometer and a temperature adjustable mounting base for the LED under study. The temperature of the mounting base was controlled using a Peltier element based temperature controlled copper plate. A temperature sensor of type PT-100 was used to measure the temperature of the mounting plate, and the current through the Peltier element was controlled to stabilize the temperature of the mounting plate. A QE65 Pro spectroradiometer from Ocean Optics was used to measure the spectral distribution of the emission intensity of the LEDs in the wavelength range between 300 nm and 1090 nm. The distance between the LEDs and the measuring head of the spectroradiometer was 150 mm. The intensity calibration of the spectroradiometer was carried out using a calibrated FEL lamp which is traceable to the spectral irradiance scale of the National Metrology Institute of Finland [6]. The wavelength scale of the spectroradiometer was calibrated against spectral lines of a mercury-argon calibration source.

Figure 1 shows measured and modelled spectra of a GaAs infrared LED and a close-up of the temperature-invariant intersection point E_B. The modeling parameters in Eq. (1) can be determined from the LED spectral measurements carried out at different temperatures. The energy E_B is located at the intersection point of the normalized spectra, and the peak shift parameter q can be determined by measuring the shift of the peak of the spectra at different temperatures. The band gap temperature shift parameter p can be found by fitting the model to the normalized spectra. For a typical infrared LED the modeling parameters are E_B = 1.54 eV, p = 4.85, and q = 4.55.

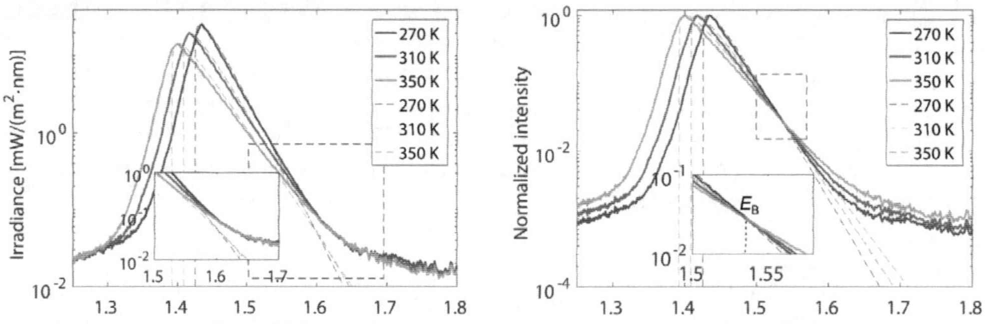

Figure 1. Measured (solid lines) and modelled (dashed lines) spectra of a GaAs LED between 270 K and 350 K and close-ups around the energy E_B. Left figure shows the absolute spectra and the right figure shows the peak normalized spectra.

As seen in Figure 1, the spectral model described by dashed lines is valid only close to the peak intensity and the intersection points [1,7]. However, despite the limitations of the model at the low-energy side of the spectrum, the model predicts reasonably well the relative intensity between the peak and the intersection energies of the normalized spectra. Figure 1 also demonstrates that the intersection point only exists in normalized spectra.

The optical characteristics such as the peak energy, intersection energy, and the emission intensity of semiconductor devices manufactured using the same process typically differ slightly between different samples. We have measured InGaN/GaN ultraviolet light emitting diodes cut from different parts of a semiconductor wafer and compared the optical properties of the samples. The LEDs were manufactured at Aalto University, grown by metalorganic chemical vapor deposition, thus the location of the samples in the wafer was known [8]. According to our studies, the standard deviation of the intersection energies is 0.007 eV if the samples are cut within 10 mm distance from each other and 0.02 eV if the samples are cut 100 mm from each other. Similar variations can be seen in the corresponding peak energies. The differences in the optical characteristics are presumed to be due to the nonideal growth process leading to crystal dislocations in the semiconductor wafer [9].

The temperature-invariant energy value E_B is a general feature of all LEDs. In Figure 1, the spectra were measured at known heat sink temperatures of 270 K, 310 K, and 350 K. However, determination of E_B does not require knowledge of the absolute temperature, but it is sufficient to vary the temperature and to determine the invariant crossing point in the relative spectra. The temperature dependent LED spectra of Eq. (1) have three modelling parameters E_B, p, and q. As seen in Eq. (2), parameter q defines the shift of the spectral peak as a function of temperature. Modeling parameter p defines the shift of the band gap energy as a function of temperature, thus after the peak shift parameter q has been determined from the measurement results, the value of p can be solved from the intensity of the intersection point using Eq. (3).

3. Reference device

An LED-based wavelength and relative intensity reference device was built utilizing the temperature-invariant energy value E_B. The device comprises a separate control unit and a temperature controlled heat sink for seven LEDs. The LEDs are attached on an aluminium heat sink equipped with a PT100 temperature sensor. The temperature value of the aluminium LED circuit board is sent to the external control unit that controls the temperature of the heat sink with a Peltier element. The temperature of the aluminium circuit board is controlled using a PI (Proportional-Integral) control algorithm. The temperature of the circuit board can be adjusted to three different values selected using a rotary switch on the front panel of the control unit. The external control unit also feeds a constant driving current of 100 mA to the LEDs and is capable of switching each LED on and off separately. Figure 2 shows a photograph of the developed LED-based reference device.

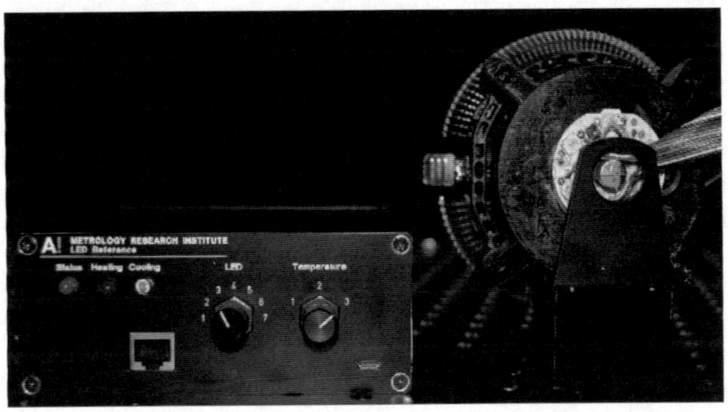

Figure 2. The LED based reference device comprises separate control and heat sink units.

The temperature-invariant reference point of each LED can be measured by varying the temperature of the heat sink and by measuring the emission spectra at different temperatures. Figure 3 shows the spectra of five different types of LEDs at three different temperatures. The measurement temperatures vary between 270 K and 398 K depending on the LED used.

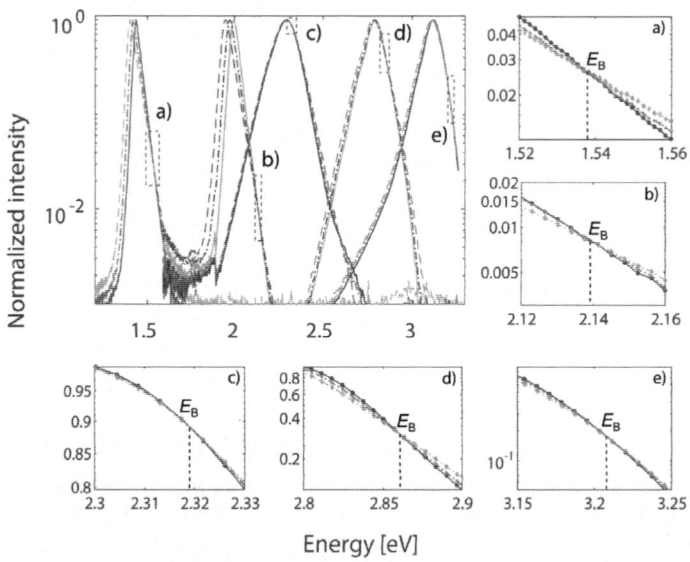

Figure 3. Normalized spectra of five LEDs measured at different temperatures. The intersection energies E_B are inside the red rectangles of dashed lines (a-e). Markers in the close-up figures indicate the measurement points.

The intersection points in Figure 3 are located at the energies (vacuum wavelengths) of 1.537 eV (806.7 nm), 2.140 eV (579.4 nm), 2.319 eV (534.7 nm), 2.860 eV (433.5 nm), and 3.209 eV (386.4 nm). The intersection energies can be used as constant reference wavelengths after calibration with a master spectroradiometer. The normalized intensities for the respective intersection points are 2.7%, 0.75%, 89.5%, 31.5%, and 17.0% of the peak intensity, and these are similarly determined by a linear master spectroradiometer.

The standard uncertainties of the energy values of the intersection points in Figure 3 are 0.005 eV, 0.008 eV, 0.003 eV, 0.005 eV, and 0.008 eV for the intersection points a)-e), respectively. The uncertainties were obtained by studying the standard deviations of the intersection points of individual pairs of curves [1]. The resolution of the spectroradiometer was 0.7 nm, leading to an energy resolution of 0.01 eV at 4.1 eV and 0.0009 eV at 1.1 eV. The standard deviation of the intensity values of the intersection points of individual pairs of curves was found to be 0.2% - 4.0% of the peak intensity.

The non-linearity of CCD (Charge-Coupled Device) array spectroradiometers has been found to be as high as 5% or more [10,11]. Excluding saturation effects, non-linearities of several per cents have been found when measuring intensity levels that are smaller than 30% of the maximum intensity [11]. We introduced an artificial non-linearity to the measurement results in Figure 3 in such a way that at the minimum intensity the non-linearity was 5% and at the maximum intensity the non-linearity was 0%, varying linearly in between. According to these analysis results, the intersection energies varied less than 0.012 eV. At low intensity intersection points a) and b) in Figure 3, the changes of the normalized intensities of the intersection points were +4.9% and +3.9% of the peak intensity, respectively. For the high intensity intersection points c)-e), the changes of the intensities of the intersection energies were +0.41%, +1.1%, and -13.6% of the peak intensity, respectively. For the ultraviolet LED e), the large change of the intensity may be mainly due to the low intensity and energy resolution. As the uncertainties of the energy and intensity values would be substantially the same as the resolution of the master spectroradiometer, the performance of a CCD array spectroradiometer could be studied with the proposed method.

4. Discussion and conclusions

Comprehensive investigation of the stability of spectroradiometers and their suitability for absolute calibrations has been carried out in [10]. Due to the aging, the drift of the wavelength scale of a spectroradiometer can be more than 1 nm and the nonlinearity can be several per cents [11]. It has been suggested that a thorough characterization is needed to verify the suitability of a spectroradiometer for absolute calibration [10]. Full verification of the linearity and the wavelength scale of a spectroradiometer will require non-portable, laboratory grade equipment. Typically the wavelength scale of a spectroradiometer is verified using pencil type gas discharge lamps with discrete emission spectral lines. As the intensities of the spectral peaks of a pencil lamp are normally not controlled, a separate facility for the linearity measurement is needed [11].

Our new type of intensity and wavelength reference can be utilized to carry out a fast verification measurement of array spectroradiometers using a single instrument only. The uncertainty of such a verification is higher as compared to the laboratory grade verification described above, but the cost is less for the cases where the slightly higher uncertainty is enough. The intersection points seen in Figure 3 can be used as temperature-invariant reference values of relative intensity and wavelength in LED spectra. It is sufficient that the LED control unit allows measurements at a few different temperatures without accurate and reproducible temperature stabilization. In addition to wavelength reference, the intersection points give information about the linearity of the spectrometer through the relative intensities at the intersection energies E_B. The wavelength uncertainty of the method is smaller than the 0.7 nm resolution of the spectroradiometer used in our test measurements. The uncertainty of the relative intensity scale was found to be 0.2% - 4.0% of the peak intensity, depending on the LED used.

Acknowledgement

The work leading to this study was funded by the European Metrology Research Programme (EMRP) projects "Metrology for III-V materials based high efficiency multi-junction solar cells" and "Towards an energy-based parameter for photovoltaic classification." The EMRP is jointly funded by the participating countries within European Association of National Metrology Institutes (EURAMET) and the European Union.

References

[1] Baumgartner H, Vaskuri A, Kärhä P and Ikonen E 2016 *Appl. Phys. Lett.* **109** 231103
[2] Schubert E F 2010 *Light-Emitting* Diodes (Cambridge: University Press)
[3] Chen N C, Wang Y N, Tseng C Y, and Yang Y K 2006 *Appl. Phys. Lett.* **89** 101114
[4] Chen N C, Lin M, Shen C, Lien W C and Lin T Y 2008 *Opt. Express* **16** 20759-73
[5] Chen N C, Lien W C, Yang Y K, Shen C, Wang Y S and Chen J F 2009 *J. Appl. Phys.* **106** 074514
[6] Kübarsepp T, Kärhä P, Manoocheri F, Nevas S, Ylianttila L and Ikonen E 2000 *Metrologia* **37** 305-312.
[7] Vaskuri A, Baumgartner H, Kärhä P, Andor G and Ikonen E 2015 *J. Appl. Phys.* **118** 203103
[8] Suihkonen S, Lang T, Svensk O, Sormunen J, Törmä P T, Sopanen M, Lipsanen H, Odnoblyudov M A and Bougrov V E 2007 *J. Crystal Growth* **300** 324-329
[9] Duxbury N, Bangert U, Shang P, Thrush E J and Jacobs K 1999 *Inst. Phys. Conf. Ser.* **161** 207-211
[10] Nevas S, Teuber A, Sperling A and Lindemann M 2012 *Metrologia* **49** S48-S52
[11] Pulli T, Nevas s, Gawhary O E, van den Berg S, Askola J, Kärhä P, Manoocheri F and Ikonen E 2017 *Appl. Opt.* **56** 3077-3086

IOP Conf. Series: Journal of Physics: Conf. Series **972** (2018) 012011 doi:10.1088/1742-6596/972/1/012011

Methods of total spectral radiant flux realization at VNIIOFI

Evgeniy Ivashin[1], Jan Lalek[2], Andrzej Rybczyński[2], Sergey Ogarev[1], Boris Khlevnoy[1], Dmitry Dobroserdov[1] and Victor Sapritsky[1]

[1]All Russian Research Institute for Optical and Physical Measurements (VNIIOFI), Moscow, Russia
[2]GL-Optics, Puszczykowo, Poland

Corresponding e-mail address: ogarev-m4@vniiofi.ru

Abstract. VNIIOFI carries out works on realization of independent methods for realization of the total spectral radiant flux (TSRF) of incoherent optical radiation sources - reference high-temperature blackbodies (BB), halogen lamps, and LED with quasi-Lambert spatial distribution of radiance. The paper describes three schemes for measuring facilities using photometers, spectroradiometers and computer-controlled high class goniometer. The paper describes different approaches for TSRF realization at the VNIIOFI National radiometric standard on the basis of high-temperature BB and LED sources, and gonio-spectroradiometer. Further, they are planned to be compared, and the use of fixed-point cells (in particular, based on the high-temperature δ(MoC)-C metal-carbon eutectic with a phase transition temperature of 2583 ^{0}C corresponding to the metrological optical "source-A") as an option instead of the BB is considered in order to enhance calibration accuracy.

1. Introduction

Accuracy of light measurements in all industries, in scientific research significantly influences the level of development of new technology, and quality of manufactured lighting products. One of the main quantities in practice is the luminous flux.

At present, in the rapidly developing technologies for the production of a new generation of light-emitting diodes (LEDs) characterized by a wide variety of shapes of emission spectra and the widespread use of modern compact high-speed spectroradiometers on CCD matrices, which allow measuring such spectral characteristics of radiation sources as color coordinates, color rendering index and correlated color temperature, in addition to measuring the integral characteristics - luminous flux and illumination value, it is extremely important and relevant to reproduce the total spectral radiant flux (TSRF) to ensure, at the present level, the unity of measurements in the field of spectroradiometry.

Nowadays VNIIOFI is proceeding with works aimed on creating spectroradiometric facilities based on a goniometer and spectroradiometer for realization and dissemination a unit of spectral radiant flux in the mode of total or partial flux from standard radiation sources in order to improve metrological support of energy-saving lighting products based on LEDs. The article discusses various methods for reproducing and transfer of TSRF on reference sources - the blackbody (BB) or calibrated LED, which are currently being studied at VNIIOFI, with the introduction of a goniometer and spectroradiometer into the standard facility, which expands its functionality as a result of the reproduction of the standard spatial distributions of radiance and irradiance of monochromatic radiation.

The TSRF quantity is being realized for the first time in the Russian Federation in the wavelength range from 0.3 μm to 1.1 μm based on a goniometer and spectroradiometer with envisaged standard uncertainty u_c from 0.19 % (at 0.65 μm) to 0.94 % (at 0.3 μm).

2. Luminous flux realization at national standards

Realization of luminous flux and TSRF of non-coherent optical radiation was carried out at VNIIOFI's national standards [1] by several methods, starting from 2003. The realization methods are based on a high-temperature wide-aperture blackbody (HTWABB) of BB3500M model as an external source of reference radiation with Planckian spectral distribution (see Fig.1) [2]. An alternative technique of luminous flux realization is based on a LED as a reference optical source, whose characteristics are measured by goniophotometer with V(λ) – corrected photometer.

The basis for the reproduction of the unit of luminous flux performed at the initial stage of work at VNIIOFI was the comparison of the total flux of the photometric lamp and the luminous flux of the HTWABB in a certain solid angle using an integrating sphere (IS), when the signal is recorded by means of photometer. The value of luminous flux

$$\Phi_l = \int_\Omega I(\theta,\varphi)d\Omega = \frac{l^2}{k_0}\int_{\Omega_0} i(\theta,\varphi)d\Omega \quad (1)$$

is determined from the measured signal $i(\theta,\varphi)$ of the photometer defined by the angles θ,φ. Here we introduce the notations: $k_0 = i_0/E_0$ is the illuminance responsivity factor, l is the distance from the radiation source to the input aperture of the IS on which the photometer is placed. The quantity of illuminance when calibrating a photometer against BB is defined as

$$E_o = 683 \cdot \int_{380}^{780} V(\lambda)E_0(\lambda,T)d\lambda \quad (2)$$

in the visible wavelength range.

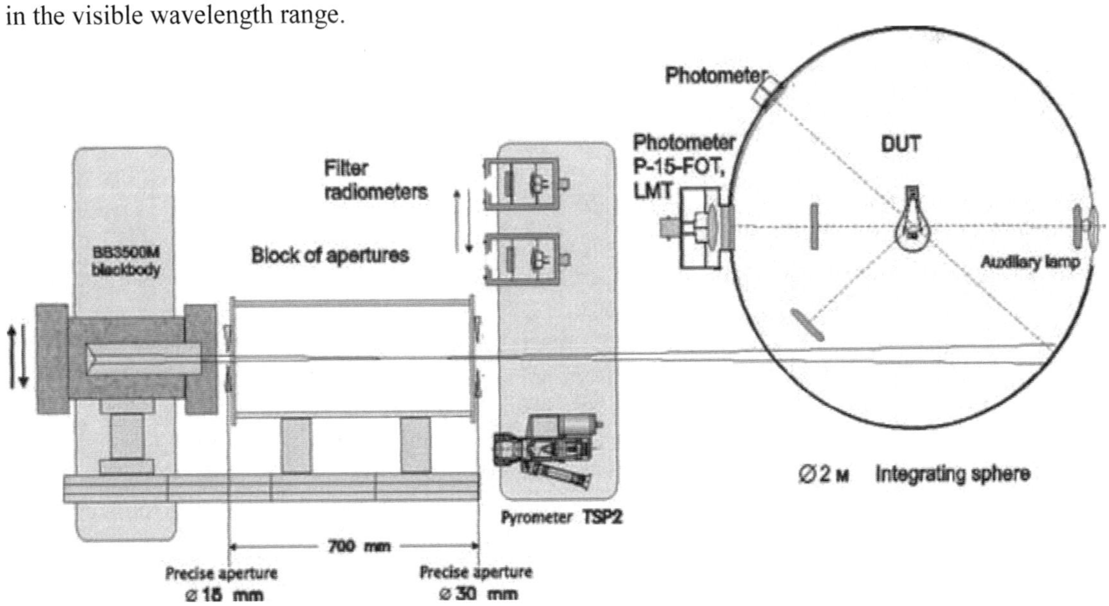

Figure 1. Scheme of realization of luminous flux, spectral radiance (SR), spectral irradiance (SI) and spectral radiant intensity (SRI) at VNIIOFI National standard based on a high-temperature wide-aperture blackbody of BB3500M model as an external source.

IOP Conf. Series: Journal of Physics: Conf. Series **972** (2018) 012011 doi:10.1088/1742-6596/972/1/012011

To realize the luminous flux quantity by means of BB and photometer, the VNIIOFI measurement facility [2] was used consisting of: BB3500M blackbody with precision diaphragms and a system for measuring its thermodynamic temperature based on absolutely calibrated filter radiometer or (alternatively) on the TSP2 pyrometer calibrated against the high temperature fixed points with assigned thermodynamic temperatures [3]; a group photometer with amplifiers and a system for measuring its spectral characteristics; IS with 2-m diameter; systems for power supply, registration and automated mechanical movement and alignment.

3. TSRF realization with spectroradiometer and photometer

Modern requirements to optical sources assume knowledge of not only the spatial, but also the spectral distribution of their radiation. The work on modernization of the National standard allowed adding into the TSRF measurement scheme a spectroradiometer (see Fig.2) and a goniometer scanning the radiation source in 2π-geometry (Fig.3). Reproduction of TSRF from an external source (either BB or LED) using a spectroradiometer as an independent measurement tool provides spectral information, in addition to a photometer:

$$\Phi_l = 683 \cdot l^2 \int_{380}^{780} V(\lambda) \iint_{0\,0}^{2\pi\,\pi} \frac{i(\lambda,\theta,\varphi)}{k_0(\lambda)} sin\theta d\theta d\varphi d\lambda \qquad (3)$$

where $i(\lambda,\theta,\varphi)$ – is the signal measured by a spectroradiometer using a goniometer, $k_0(\lambda)$ - is the spectral illuminance responsivity factor.

The same reference source with spectral radiant intensity $I(\theta,\varphi)$ is measured with a photometer and $I(\lambda,\theta,\varphi)$ with a spectroradiometer, and their luminous fluxes can be equated:

$$683 \cdot \int_{380}^{780} V(\lambda) i(\lambda) d\Omega d\lambda \equiv \int_{\Omega_0} I d\Omega = \Phi_l. \qquad (4)$$

Relative measurements of the signal $i(\lambda)$ are carried out by a non-calibrated spectroradiometer, and we find as follows:

$$I_{rel}(\lambda) = \frac{i(\lambda)}{i(\lambda_0)} = \frac{I(\lambda)}{k_{scale}(\lambda)}. \qquad (5)$$

As a result, we find the required TSRF

$$\Phi(\lambda) = \int_{380}^{780} k_{scale}(\lambda) \cdot I_{rel}(\lambda) d\lambda. \qquad (6)$$

The upgraded scheme of realization of the TSRF from optical source (in comparison of the radiation fluxes from an external source, BB or LED, and from the DUT under calibration, for example a lamp located in the center of the IS) using the spectroradiometer is shown in Fig.2.

The measuring facility includes a 2-m integrating sphere (with $BaSO_4$ coating and 50-mm entrance aperture), a spectroradiometer of the Konica Minolta CS-2000 type (with diffraction grating and Hamamatsu back thinned CCD working in a range of 380-780 nm, physical resolution approx. 0.5 nm, optical resolution FWMH approx. 3.5 nm, integration time 10-10000 ms); fast photometer (Hamamatsu photodiode with V(λ)-correction in class A < 1.5 % and < 0.1 % linearity, equipped with low-noise electronics of a dynamic range of 6 orders of magnitude and speed of 50 measurements per second), and mounting unit.

LEDs with the following characteristics are used as a quasi-Lambert source for the spatial distribution of the source brightness: power 30 W, luminous flux 1500 lm, color temperature 3500 K. The spectral flux of the LED source will be measured by a gonio-spectroradiometer in 2π-geometry.

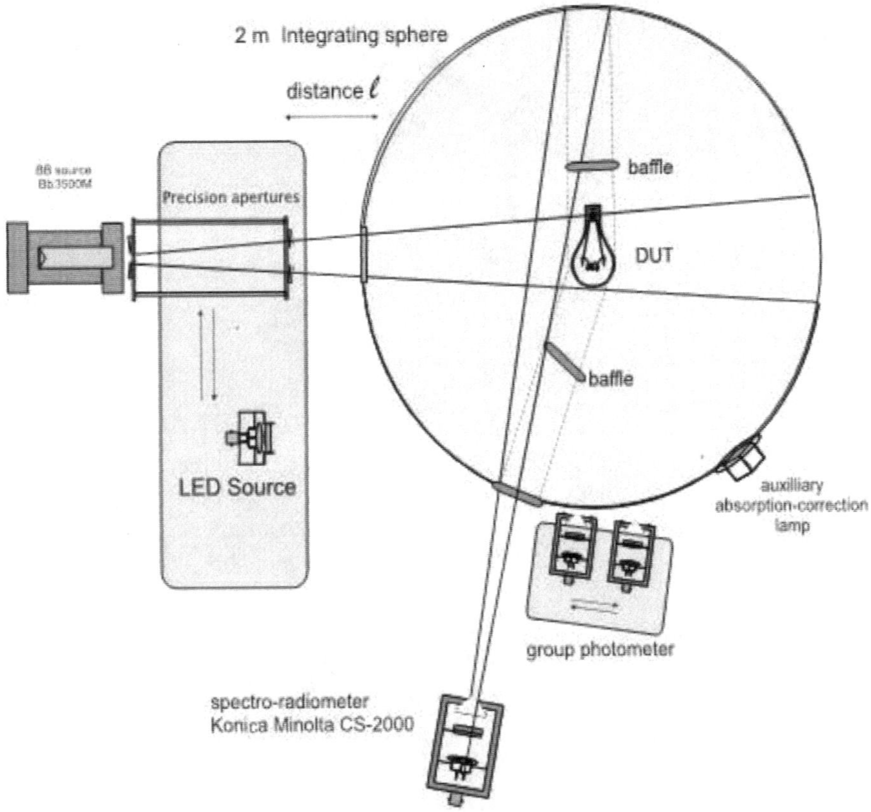

Figure 2. Scheme of TSRF realization applying BB or LED as an external source.

4. TSRF realization on the basis of gonio-spectroradiometer

From the analysis of the results of patent studies of practically used goniometers and spectroradiometers for measuring the optical flux of optical sources, it has been established that the most suitable and popular for the study of the characteristics of photometric lamps and light-emitting diodes are structures that ensure the angular rotation of the source in the horizontal plane while simultaneously scanning the goniometer's movable lever at other angles with a detector node fixed on it. Measuring devices that provide such a scheme for realization and dissemination of the TSRF unit that corresponds to the R&D chosen at this stage for constructing the VNIIOFI standard facility use the designs developed in NIST (USA) [4], PTB (Germany) [5], NMIJ (Japan) [6] and LMT Lichtmesstechnik GmbH Berlin (Germany) [7].

The realization of the TSRF based on a reference source – the blackbody radiator or the calibrated LED source, when the goniometer and the spectroradiometer are included in the standard, extends its functionality as a result of the realization of the reference spatial distributions of the irradiance and radiance for monochromatic radiation. In addition, the novelty of the works performed on the standard is the realization of an additional, radiometric value, the spectral density of the total radiant flux, which is extremely important to unify measurements in the field of spectroradiometry at the present level by transferring units to organizations utilizing secondary standards, centers of standardization and metrology, laboratories, carrying out measurements of spectroradiometric and photometric characteristics of products of energy-saving LEDs, and multielement IR radiation detectors.

The third scheme for the realization of the TSRF from optical source in the form of a halogen lamp is based on a goniometer (made by GL Optic, Poland) (Fig. 3), which includes a 127-mm integrating sphere with a spectroradiometer (1) and a high-speed photometer class A (Fig. 4) - all elements are temperature stabilized.

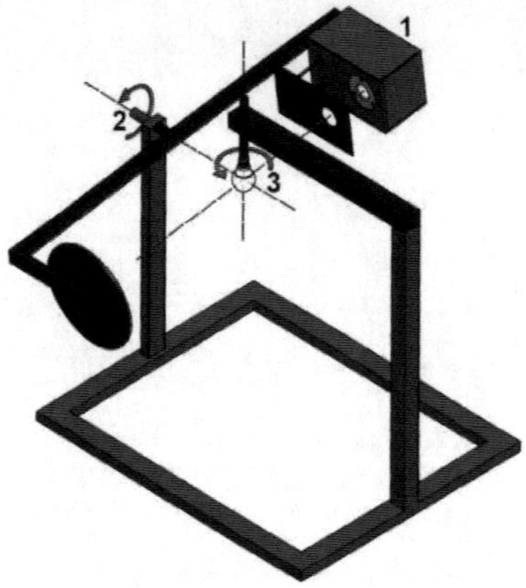

Figure 3. Computer-controlled high-class goniometer designed and manufactured by GL-Optics. 1) Integrating sphere with temperature-stabilized spectroradiometer and high-speed photometer; 2) Main rotation axis - uniform rotation with a constant angular speed; 3) DUT slow uniform rotation with a constant angular speed.

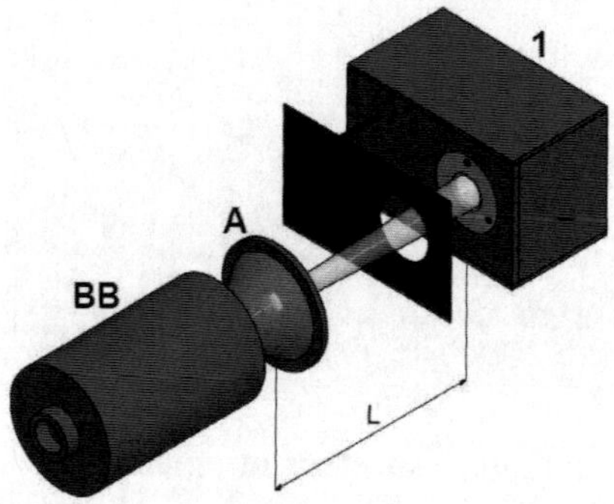

Figure 4. Scheme of calibration of the spectroradiometer and photometer located in the detector block (1), according to spectral characteristics, according to the reference BB with the radiation flux generator in the form of a block of precision apertures (A), similar to the scheme in Fig. 1.

Characteristics of computer-controlled high-class goniometer designed on the basis of blackened aluminum profiles are as follows: radius of measuring arm 500 mm, both axis driven be a stepping motor with a smooth acceleration and speed control, set of baffles reducing stray light, handle for precise fix of light sources (170-mm max. size of the light source), possibility of changing a position of light source within 180 degrees, laser system for geometry control. Rotation axis (2) and (3) of uniform rotation with a constant angular speed are shown in the figure. Mounting unit enables precise fixation of a measuring unit to a goniometer arm, or to a calibration system based on BB source.

The calibration scheme of the spectroradiometer and photometer according to the spectral characteristics of the reference blackbody radiator with a block of precision apertures (A), similar to the diagram in Fig. 1, is shown in Fig. 4. In this case, the radiation from the BB falls on the input of the integrating sphere, at the output of which a spectroradiometer and a photometer are located.

5. VNIIOFI project on developing high power LED light sources

VNIIOFI has started a project on developing and investigating high power LED light sources, which are to be used as standard sources for luminous intensity, luminous intensity distribution, luminous flux and colour measurements. Design of LED source (presented in Fig.5) features a light head and a parametric current source (PCS). Three heads were built: two white with power of 30 W and 68 W and a blue one of 30 W. Each LED was selected as the most stable from a batch after seasoning (with photometric stability of 0.02 % for approximately 10 h) [8].

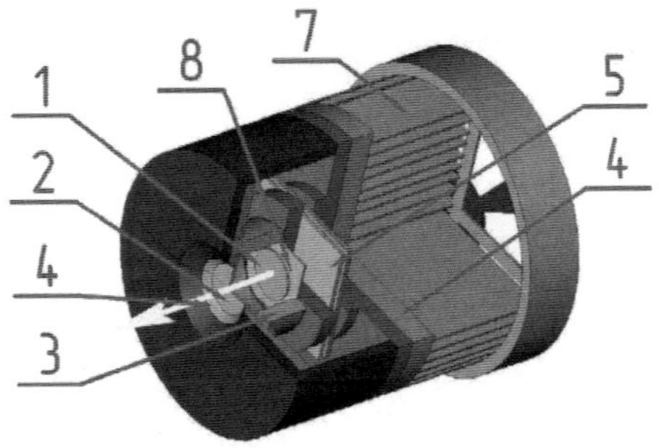

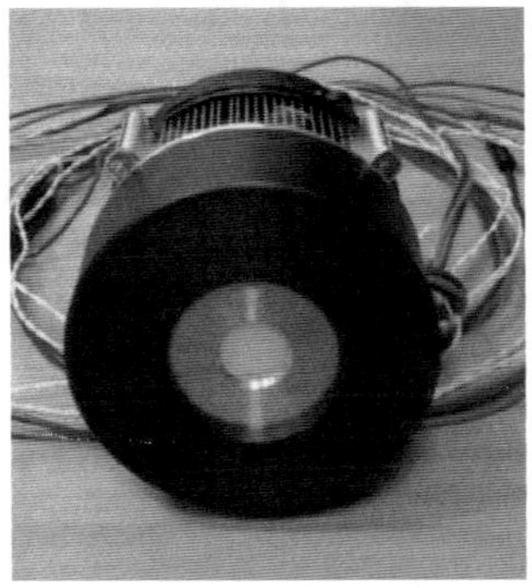

Figure 5. Scheme and external view of a standard LED source designed at VNIIOFI.
1) Chip-on-board type LED, 2) Opal glass, 3) PTFE tube, 4) Aperture, 5) Peltier element, 6) Heatsink, 7) Fan, 8) Pt1000 temperature sensor.

6. Conclusions

The paper describes different approaches for TSRF realization at the VNIIOFI National radiometric standard on the basis of high-temperature BB and LED sources, and gonio-spectroradiometer. The TSRF quantity is realized for the first time in the Russian Federation in the wavelength range from 0.3 μm to 1.1 μm based on a goniometer and spectroradiometer with envisaged standard uncertainty u_c from 0.19 % (at 0.65 μm) to 0.94 % (at 0.3 μm).

Further, the use of fixed-point cells [9] (in particular, based on the high-temperature δ(MoC)-C metal-carbon eutectic with a phase transition temperature of 2583 ^{0}C corresponding to the metrological optical "source-A") as an option instead of the BB is considered in order to exclude temperature measurements with filter radiometer or TSP-2 for enhancing calibration accuracy.

Acknowledgments

The work was carried out using the equipment of the Center for the collective use of high-precision measuring technologies in the field of photonics (ckp.vniiofi.ru), created on the basis of FSUE VNIIOFI and supported by the Ministry of Education and Science of Russia in the framework of the implementation of agreement No. 14.595.21.0003 of August 28, 2017 (the unique identifier RFMEFI59517X0003).

References

[1] http://www.vniiofi.ru/etalon.html
[2] Ogarev S A, Sapritskii V I, et al. *Contemporary state of providing uniformity of luminous quantity measurements* 2005 *Meas.Tech*, **48**, No.11, pp.1090-1095. doi: 10.1007/s11018-006-0025-8
[3] D.H. Lowe et al. *The equilibrium liquidus temperatures of rhenium–carbon, platinum–carbon and cobalt–carbon eutectic alloys*, 2017 *METROLOGIA*, **54**, No.3, S.390–8
[4] http://www.nist.gov
[5] http://www.ptb.de
[6] http://www.nmij.jp
[7] http://www.lmt-berlin.de
[8] Ivashin E., Ogarev S., et al. *High power LED standard light sources for photometric applications.* Accepted for publ. at the *NEWRAD special issue IOP conference proceedings*
[9] Sapristky V I, Ogarev S A, Khlevnoy B B, Samoylov M L, Khromchenko V B *Development of metal–carbon high-temperature fixed-point blackbodies for precision photometry and radiometry* 2003 *METROLOGIA*, **40**, No.1, S128-S131

Key comparison CCPR-K1.a as an interlaboratory comparison of correlated color temperature

P Kärhä[1], A Vaskuri[1], T Pulli[1] and E Ikonen[1,2]

[1] Metrology Research Institute, Aalto University, Espoo, Finland
[2] MIKES Metrology, VTT Technical Research Centre of Finland Ltd, Espoo, Finland

E-mail: petri.karha@aalto.fi

Abstract. We analyze the results of spectral irradiance key comparison CCPR-K1.a for correlated color temperature (CCT). For four participants out of 13, the uncertainties of CCT, calculated using traditional methods, not accounting for correlations, would be too small. The reason for the failure of traditional uncertainty calculation is spectral correlations, producing systematic deviations of the same sign over certain wavelength regions. The results highlight the importance of accounting for such correlations when calculating uncertainties of spectrally integrated quantities.

1. Introduction

Spectral irradiance measurements of National Metrology Institutes (NMI) were compared in the CCPR Key Comparison K1.a [1], a measurement campaign that took place during years 2001-2003. Correlated Color Temperature CCT [2] is a quantity often derived from measured spectra, but there has not been a key comparison of CCT. We analyzed the data of CCPR-K1.a to calculate CCTs that the participants would obtain for a lamp that resembles the CIE Standard Illuminant A, and conventional uncertainties for the derived CCTs [3]. We also discuss the effects of correlated deviations of spectral irradiance values over the visible region to explain the results on CCT and its uncertainty.

2. Analysis of deviations

The report of CCPR-K1.a lists for each participant, at various wavelengths, the Degree of Equivalence (DoE) which is essentially the deviation of the spectral irradiance values from the world mean, and the uncertainty of the DoE consisting of participant's uncertainty and the comparison uncertainty. The data were interpolated within the approximate visible region (350 nm – 850 nm) at 1 nm interval to allow calculation of the CCT as shown in figure 1.

The interpolation was carried out with N^{th} degree polynomials, where N was approximated visually for each data set. This procedure effectively separates random noise in measurements from structural deviations caused by spectral correlations. The structural deviations were further divided to offset and higher order structural deviations. Table 1 shows how the deviations of the participants separated into these three classes. Offset was calculated as the average of the spectral deviations, (random) noise as the standard deviation of the residual errors after subtracting the interpolated values, and the structural deviation was calculated as the standard deviation of the interpolation function. The key comparison reference values (KCRV) may also contain spectral structure as all NMIs did not correct their results with respect to correlations. However, there are various shapes of spectral structures present in figure 1 which could not all be explained by spectral structure in the KCRV.

IOP Conf. Series: Journal of Physics: Conf. Series **972** (2018) 012012 doi:10.1088/1742-6596/972/1/012012

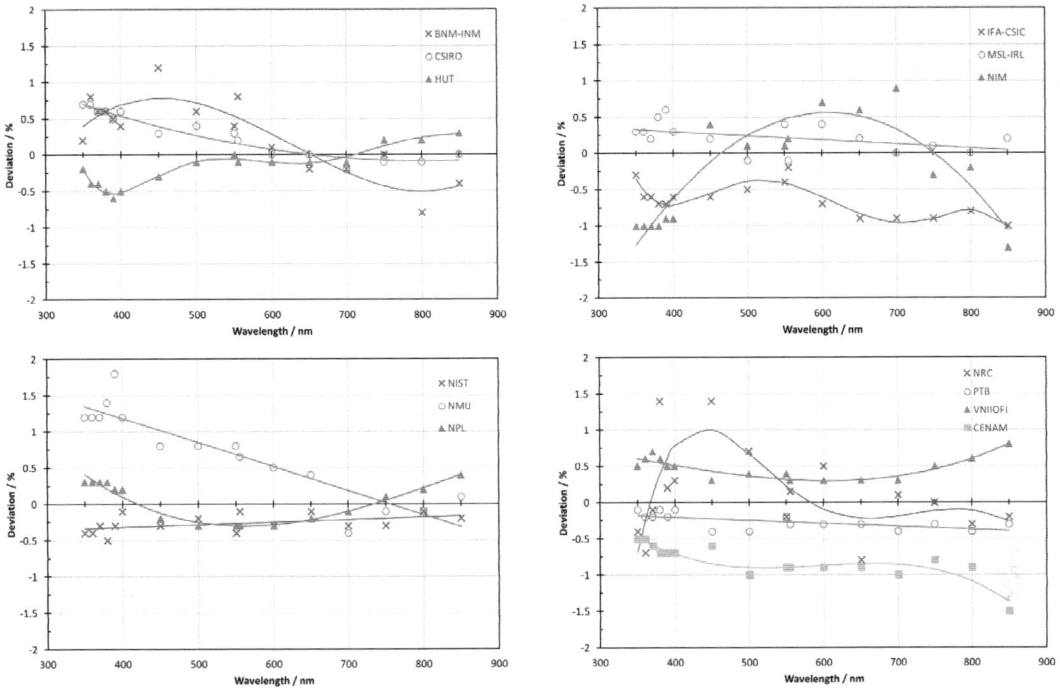

Figure 1. Interpolated deviations from CCPR-K1.a reference value (unilateral DoE) for participating institutes. The abbreviations of laboratory names are defined in table 1.

Table 1. Distribution of spectral deviations to different structural classes, and the degree N used for the interpolation polynomial.

			Deviation / %			
Laboratory	Country	Abbreviation	Noise	Offset	Structural	N
Bureau National de Métrologie - Institut National de Métrologie	France	BNM-INM	0.24	0.29	0.45	3
Commonwealth Scientific & Industrial Research Organisation	Australia	CSIRO	0.08	0.28	0.30	2
Helsinki University of Technology (Aalto University)	Finland	HUT	0.04	0.17	0.26	6
Instituto de Física Aplicada - Consejo Superior de Investigaciones Científicas	Spain	IFA-CSIC	0.08	0.17	0.26	5
Measurement Standards Laboratory of New Zealand - Industrial Research Limited (Callaghan Innovation)	New Zealand	MSL-IRL	0.18	0.22	0.10	1
National Institute of Metrology	China	NIM	0.29	0.29	0.67	2
National Institute of Standards and Technology	USA	NIST	0.11	0.27	0.06	1
National Metrology Institute of Japan	Japan	NMIJ	0.26	0.72	0.56	1
National Physical Laboratory	UK	NPL	0.06	0.04	0.26	3
National Research Council	Canada	NRC	0.44	0.13	0.46	5
Physikalisch-Technische Bundesanstalt	Germany	PTB	0.08	0.26	0.07	1
All-Russian Scientific Research Institute for Optical and Physical Measurements	Russia	VNIIOFI	0.07	0.48	0.14	3
Centro Nacional de Metrología	Mexico	CENAM	0.11	0.82	0.22	3
Average deviation			0.16	0.32	0.29	
Fraction of total deviation			21 %	41 %	38 %	

3. Correlated colour temperature

The interpolated deviations of spectral irradiance were used to calculate deviations that the participants would see in the CCT of a lamp resembling the CIE Standard Illuminant A. The spectrum of the CIE Standard Illuminant A was disturbed with the deviation function, and the calculated CCT was compared to 2856 K to obtain ΔCCT. The results are shown in table 2. The uncertainties for DoE were wavelength dependent. To simplify calculations, an average uncertainty over the visible region was used.

As can be seen in table 2, for four of the laboratories ΔCCT would exceed U(CCT). Statistically it would be expected that there are no or at most one participant exceeding $|E_n| = 1$. In spectral irradiance, there was an agreement within coverage factor $k = 2$ in the wavelength range 290 nm – 800 nm for all participants. It is worth noting that the average of the ΔCCT values in table 2 is positive, which in an intercomparison would most likely be masked by the method of calculating the reference value. If we introduce an offset of -0.7 K to the deviations, i.e. the average of the ΔCCTs weighted with U^{-2}, the E_n values would change, but there would still be four laboratories with $|E_n| \geq 1.2$.

Table 2. Expanded uncertainties of spectral irradiance (DoE), deviations of CCT, calculated expanded uncertainties of CCT [3], and E_n values. The E_n = ΔCCT / U(CCT) indicates whether the deviation is within expanded uncertainty.

Laboratory	U(DoE)	ΔCCT/K	U(CCT)/K	E_n
BNM-INM	2.0 %	8.6	5.6	1.5
CSIRO	0.8 %	3.4	2.2	1.5
Aalto (HUT)	0.8 %	0.4	2.2	0.2
IFA-CSIC	3.4 %	5.5	9.5	0.6
MSL-IRL	1.5 %	1.1	4.2	0.3
NIM	1.0 %	-3.8	2.9	-1.3
NIST	0.9 %	-0.6	2.5	-0.3
NMIJ	2.7 %	6.2	7.6	0.8
NPL	0.6 %	-0.2	1.7	-0.1
NRC	2.9 %	11.3	8.1	1.4
PTB	1.1 %	0.8	3.1	0.3
VNIIOFI	0.9 %	0.8	2.5	0.3
CENAM	4.0 %	-0.6	11.2	-0.1

4. Discussion

It is quite evident that although there was a good agreement within $k = 2$ in the spectral irradiance intercomparison, the participating laboratories would have deviations outside their expanded uncertainties in a CCT intercomparison, if the uncertainties were analyzed by conventional methods not accounting for correlations. We conclude that this is due to systematic structures in the spectral irradiance deviations as shown in figure 1 and table 1. These structures are introduced by correlations in the uncertainties of spectral irradiance values at different wavelengths [4] that should be taken into account in the uncertainty analysis. Sources for the structures and the related correlations may include interpolation of data that introduces complicated structures in the deviations, temperature determination of black body radiators that introduce slope-like errors in the data, and geometrical factors that introduce offsets in the data. If known, these correlations should be accounted for in the uncertainty analysis.

Comparison CCPR-K1.a was never intended to be a CCT intercomparison. Therefore, one should not pay attention to the E_n-values of any particular laboratory. Most laboratories could take the correlations into account and the level of agreement would thus be better. In this sense, this paper highlights the importance of arranging an intercomparison of CCT or including it in the analysis of a key comparison of spectral irradiance.

For laboratories not making scale realization themselves, but rather taking traceability from an NMI, the correlation data are rarely available. Thus, when these laboratories measure CCT with their calibrated spectroradiometers, there is a danger that their uncertainty estimates may be too small. In such cases, the fractions calculated at the end of table 1 may be useful. On the average, the uncertainties of the spectral irradiance values of NMIs in the visible region may be assumed to be 20% uncorrelated (noise), 40% offset, and 40% structural. The effect of noise averages out in the spectral integration involved in the CCT calculation, offset does not affect CCT at all, but the structural uncertainty is what causes largest measurement errors. When using a purchased lamp for calibration, the noise will increase. Thus fractions of 1/3, 1/3 and 1/3 (for variances) for the introduced correlation classes could well be used to characterize a calibrated spectroradiometer.

It is complicated to take correlations into account in uncertainty analysis if they are not known. Thus, they are often neglected. We have developed a new Monte Carlo based method that can be used to study effects of possible correlations [4]. The method can be used to calculate uncertainties assuming no correlations, full correlation (offset) and severe correlations assuming worst possible scenario. Using the fractions derived above, the method can also be used to derive uncertainties assuming typical correlation scenarios. Our Monte Carlo method [4] gives higher expanded uncertainties for CCT that would cover the noted errors arising from the structural spectral deviations in table 1. We have also applied the method for total ozone column measurements of solar UV [5, 6].

Acknowledgments
This work has been supported by the European Metrology Research Programme (EMRP) within the joint research project ENV59 "Traceability for atmospheric total column ozone" (ATMOZ). The EMRP is jointly funded by the EMRP participating countries within EURAMET and the European Union.

References

[1] Woolliams E R, Fox N P, Cox M G, Harris P M and Harrison N J 2006 Final report on CCPR K1-a: Spectral irradiance from 250 nm to 2500 nm *Metrologia* **43** Technical Supplement 02003
[2] CIE 15.3:2004, Technical Report Colorimetry (Vienna: Comission Internationale de L'Eclairage)
[3] Gardner J L 2000 Correlated colour temperature - uncertainty and estimation *Metrologia* **37** 381–384
[4] Kärhä P, Vaskuri A, Mäntynen H, Mikkonen N and Ikonen E 2017 Method for estimating effects of unknown correlations in spectral irradiance data on uncertainties of spectrally integrated colorimetric quantities *Metrologia* **54** 524–534
[5] Kärhä P, Vaskuri A, Gröbner J, Egli L and Ikonen E 2017 Monte Carlo analysis of uncertainty of total atmospheric ozone derived from measured spectra *AIP Conference Proceedings* **1810** 110005
[6] Vaskuri A, Kärhä P, Egli L, Gröbner J and Ikonen E (submitted) Monte Carlo method for determining uncertainty of total ozone derived from direct solar irradiance spectra: Application to Izaña results

IOP Conf. Series: Journal of Physics: Conf. Series **972** (2018) 012013 doi:10.1088/1742-6596/972/1/012013

Traceability of laser pulse energy measurements by linking reference standards for CW and pulsed measurements

Özcan BAZKIR[1], Seval CENK[1], Khaled Mahmoud[2,3]

[1]*National Metrology Institute (TUBITAK-UME), P.O.Box.54, 41470 Kocaeli/Tukey*

ozcan.bazkir@tubitak.gov.tr

seval.cenk@tubitak.gov.tr

[2]*National Measurement and Calibration Center (NMCC) Riyadh/Saudi Arabia*

[3]*National Institute for Standards (NIS), PO Box; 136 Giza, Code No 12211, Egypt*

k.abdelftah@saso.gov.sa

Corresponding: ozcan.bazkir@tubitak.gov.tr

Abstract. A method linking pulse energy scale to continuous wave (CW) power scale is described. The connection between CW power and pulse energy is established by generating pulses from CW beam, and correlating the CW signal measured by reference optical power meter and pulsed-signal measured by calorimeter. In this connection CW laser source from which pulses are generated is Nd:YAG laser and the lengths of pulses adjusted are at the order of milliseconds. The realized energy scale is transferred from calorimeter to laser energy meters through using pulsed Nd:YAG laser having one fundamental (at 1064 nm) and two harmonics (at 532 nm and at 365 nm) with nano-second level pulse length.

1. Motivation

The most basic method of checking the performance of a laser is to measure accurately its power and energy. Accurate measurements of laser power and energy is key factor in making the proper decisions to be made for choosing the best equipment for the job in variety of medical, scientific, industrial and biomedical applications [1,2]. This makes it so important to properly select the suitable sensors, measurement procedures and achieve the required pulse energy measurement traceability for securing the reference measurements in the national metrology laboratories [3-6].

2. Principle and Instrument

This work constitutes two stages; first stage covers traceability of laser energy calorimeter, which is used as transfer standards for Q switch pulse laser energy measurements to the continuous wave (CW) laser power reference measurement standards, second stage covers the transfer of energy scale from laser energy calorimeter to laser energy measurement sensors.

At first stage, for the traceability of laser energy calorimeter CW Nd: YAG laser (1064 nm and 532 nm), optical power reference measurement standards (Electrically Calibrated Pyroelectric Radiometer (ECPR) and Thermal Power Meter (3A-P-SH), Scienntech Astral Series S laser energy calorimeter (AC2501S), 500 MHz oscilloscope, fast Si photodiode (UPD-50-SP, Rise time <40ps) , high speed chopper (SR540, 4kHz) and some opto-mechanical accessories were used as shown in Figure 1

The work in this section was basically carried out according to the following principle: Using a high speed optical chopper CW laser beam converted to quasi pulse beams having pulse widths of order of milliseconds. Then, this beam was passed through a beam splitter such that one part of it sent to the CW reference optical power measurement standard and another part of it was sent onto the fast photodiode. Since these measurements were carried out up to 800 mW optical power levels, in order to avoid the surface damage of measurement sensors, beam was homogenized and its diameter was expanded from 6 mm to 12 mm using laser beam expander.

Similarly on the fast photodiode branch using neutral density filters (Optical Density from 0.1OD to 4.0OD) the beam power was attenuated to below the saturation level of high speed photodiode. Apart from this beam branch, all the other laser beams emerged from the beam splitter were blocked using beam traps which are suitable for CW/Pulse laser beams up to 50 W.

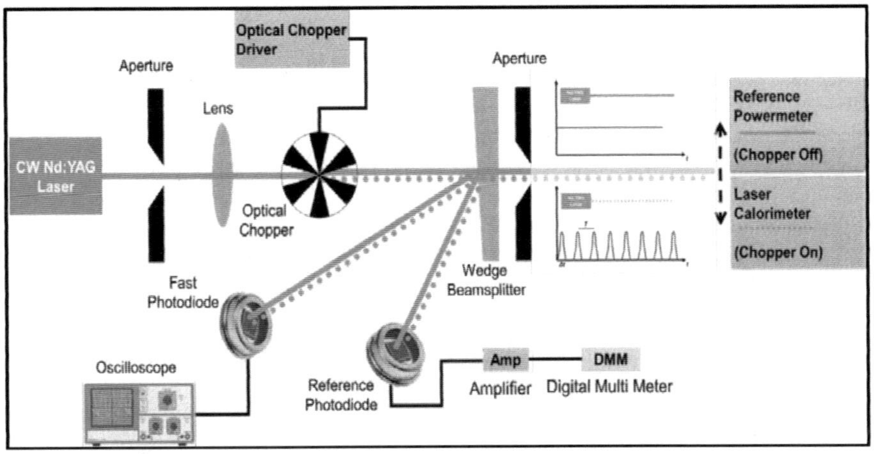

Figure 1. Measurement set up for linking pulse energy to CW power

Following to these preparation processes, initially optical chopper was turned off and the laser beam along the power measurement line was dropped on the CW laser power reference measurement standard and the optical power was recorded. Then the CW laser power reference measurement standard was substituted by laser energy calorimeter and the laser chopper turned on. The optical power recorded by laser energy calorimeter is the average optical power of quasi pulse beams created by high speed optical chopper. The peak power corresponding to this average power is the power which was measured by the CW laser power reference measurement standard. To convert this average optical power to the peak power and also to calculate energy correspond to each pulse, quasi pulse widths and laser injection time parameters recorded by high speed photodiode and oscilloscope were used, see Figure 2. Pulse energy, peak and average power for the created quasi pulse were calculated using the relations given in Eqn1.

$$P_p = \frac{E}{\Delta t} \quad \text{and} \quad P_{av} = \frac{E}{T} \qquad (1)$$

Where Pp is the peak power, Pav average power, E is the energy per pulse, Δt is the pulse width and T is the period of pulses.

The calibration factor for laser energy calorimeter was derived by adjusting its calibration factor such that to get the same power value as measured by the reference power meter. The measurements were done using two different reference optical power standards. From 10 mW to < 100 mW optical power interval ECPR which is traceable to silicon photodiode based trap detector and from 100 mW to 800 mW optical power interval 3A-P-SH which is traceable to ECPR were used. The uncertainties corresponding to these intervals were evaluated as 4,8 % and 5,2 % respectively.

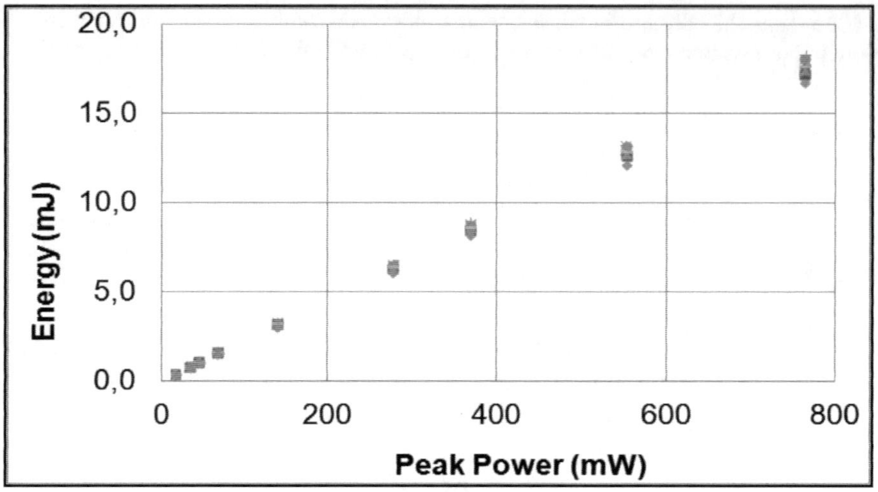

Figure 2. Results of linking the pulse energy to cw power using laser calorimeter

At second stage, for the transfer of energy scale from laser energy calorimeter to laser energy meter, Q switch Nd:YAG laser, laser energy calorimeter, laser energy meter to be calibrated which has a pyroelectric based energy sensor (J-50MB-YAG), and disply, 500 Mhz oscilloscope, fast photodiode and some opto-mechanical accessories were used, the schematic diagram of our setup is shown in Figure 3.

The J-50MB-YAG sensor which is designed for use with pulse lasers operating at 1064 nm, 532 nm and 355 nm wavelengths and with low repetition rates . The Q switch Nd:YAG laser has pulse energies of 400 mJ, 160 mJ and 110 mJ at 1064 nm, 532 nm and 355 nm wavelengths respectively. The pulse has approximately 6 ns pulse width, 20 Hz repetition rate and about 3 mm beam diameter. The above energies for Q switch Nd:YAG laser at 1064 nm, 532 nm and 355 nm wavelengths are the maximum energies and can be tuned to a few mJ level by adjusting related attenuators.

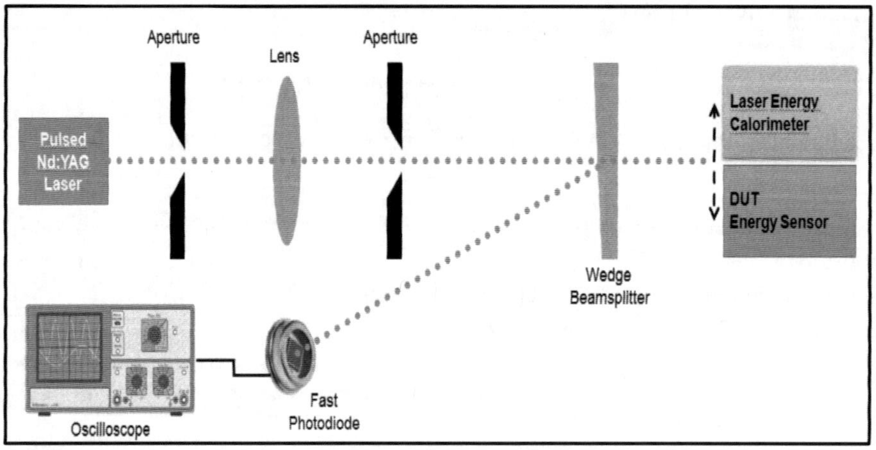

Figure 3. Measurement set up for transferring the laser energy scale from laser calorimeter to the laser energy meters

At this stage Q switch Nd:YAG laser at each wavelengths using attenuators were adjusted to their lowest energy levels and the beam emerged from the laser was directed to beam splitter. Where one

part of the beam was used for the energy measurements using laser energy calorimeter (reference) and laser energy meter and another part of the beam was used for determining the optical characteristics of pulse beam using high speed photodiode and oscilloscope.
On the energy measurement side in order to prevent surface damage on measurement sensors due to high current densities of laser beams, initially laser beam size were adjusted to about 12 mm using suitable beam expander for each wavelengths. Then damage threshold cards were put in front of sensors and from attenuators laser energy level gradually increased so as to determine maximum were respectively placed in front of pulse laser beam and their responses to energy values were recorded. The laser energy calorimeter which is used as reference device has time constant of about 3 sec. So it was adjusted to record the average power within defined injection period. Then using injection period and repetition rate of pulse, corresponding energy values was calculated for the energy range from 10 mJ to 180 mJ as presented in Figure 4. Then the calibration factor of the laser energy meter was derived by adjusting its calibration factor to get the same energy values as measured by the laser energy calorimeter.

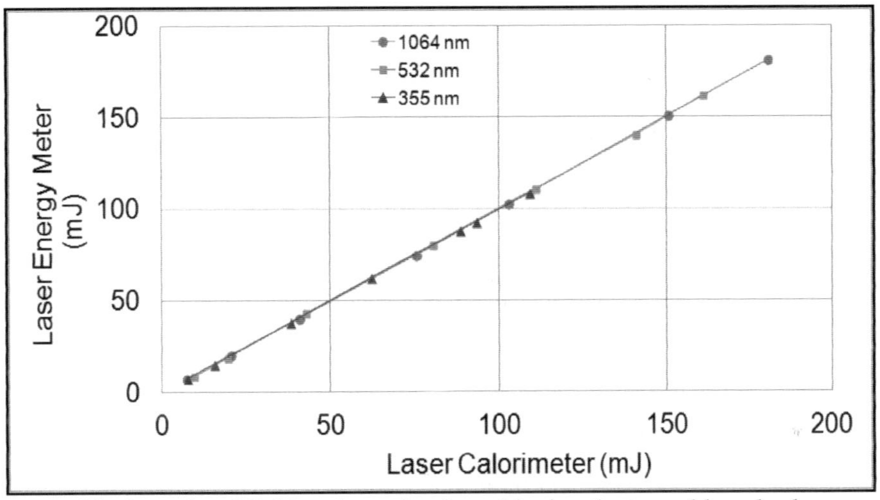

Figure 4. Final results of linking the scales for achieving the traceable pulse laser energy measurements

References

[1] A Joshua Hadler, CW Laser Power and Energy Calibrations at NIST, NIST Sp. Pub. 250-75 (2007).
[2] Marla Dowell, Pulsed-Laser Metrology at NIST, Optics and Photonic News, Vol. 12, Issue 2: 30-33 (2001).
[3] E.D. West et al., A reference Calorimeter for Laser Energy Measurements, NBS Journal of Res, A Physics and Chem 76A:13-26 (1972).
[4] E.D. West et al., Theory of Isoperibol Calorimetry for Laser Power and Energy Measurements, Journal of Applied Physics 41(6): 178-180 (1970)
[5] M G White, R Leonhardt, D Livigni and J H Lehman, A CW calibrated laser pulse energy meter for the range 1 pJ to 100mJ, Metrologia 51 (2014) 225–234
[6] Möstl K and Brandt, F A New Method to Establish a Scale for Laser Pulse Energies, Metrologia Volume 28 Number 3,1991,121-124

IOP Conf. Series: Journal of Physics: Conf. Series **972** (2018) 012014 doi:10.1088/1742-6596/972/1/012014

The development of the advanced cryogenic radiometer facility at NRC

A Gamouras, A D W Todd, É Côté, and N L Rowell

National Research Council Canada, 1200 Montreal Road, Ottawa, Ontario, Canada

E-mail: angela.gamouras@nrc-cnrc.gc.ca

Abstract. The National Research Council (NRC) of Canada has established a next generation facility for the primary realization of optical radiant power. The main feature of this facility is a new cryogenic electrical substitution radiometer with a closed-cycle helium cryocooler. A monochromator-based approach allows for detector calibrations at any desired wavelength. A custom-designed motion apparatus includes two transfer standard radiometer mounting ports which has increased our measurement capability by allowing the calibration of two photodetectors in one measurement cycle. Measurement uncertainties have been improved through several upgrades, including newly designed and constructed transimpedance amplifiers for the transfer standard radiometers, and a higher power broadband light source. The most significant improvements in uncertainty arise from the enhanced characteristics of the new cryogenic radiometer including its higher cavity absorptance and reduced non-equivalence effects.

1. Introduction

Throughout the last few decades, the National Research Council (NRC) has continued to improve its primary scales in optical radiant power and spectral responsivity. Continuing with a detector-based realization, NRC transitioned from room-temperature electrical substitution absolute radiometers to a cryogenic absolute radiometer for most of the wavelength range in 1994 [1]. Now, two decades later, this cryogenic radiometer is at the end of its service life and is being replaced with an advanced radiometry facility equipped with a new, state-of-the-art cryogenic radiometer, broadband light sources and customized apparatus for increased measurement capability. Details of the new apparatus and the impact on the scale realization uncertainties from 300 nm to 1000 nm are presented.

2. Cryogenic radiometer apparatus

The configuration of NRC's new advanced radiometry facility is shown in Figure 1. The cryogenic radiometer (CryoRad III Radiometer: L-1 Standards and Technology, Inc.) is designed to accept optical radiation from the output of a monochromator illuminated with a broad-band light source. In this new facility, a laser-driven light source (LDLS) (EQ-99X: Energetiq Technology, Inc.) and a tungsten lamp were implemented to cover the spectral range of 250 nm to 2700 nm. These broad-band light sources were mounted on a linear translation stage at the monochromator entrance allowing for rapid and repeatable alignment to the monochromator entrance slit. Two concave spherical mirrors focused light from the output of a double-subtractive monochromator, through a wavelength order sorting filter wheel to a tilted optical window mounted on a flexible vacuum bellows assembly. This large vacuum assembly was mounted on a linear stage with a pin and pivot system to rotate the cryogenic radiometer and transfer standard detectors for alignment to the common optical path (30° motion). The NRC transfer standard radiometers were designed and constructed with an integrated front window [1] which provided a vacuum seal when the radiometers were mounted to the evacuated vacuum bellows assembly. Isolation valves between the bellows vacuum chamber and the transfer radiometer mounting ports allowed the detectors to be changed during system operation. Fine adjustment of the transfer detector alignment was achieved using individual translation stages. To minimize the effect of any light source instabilities, transfer radiometer measurements were bracketed by cryogenic radiometer measurements. The radiant flux was also monitored using a separate detector that measures the reflection from the optical window.

To minimize any difference in the optical power incident to the cryogenic radiometer cavity and the transfer standard detectors, a circular aperture was mounted at the monochromator output resulting in a focused beam spot diameter of 5 mm. This spot diameter under-filled the cryogenic radiometer cavity input (7 mm diameter) as well as the input to the NRC transfer standard radiometers (6 mm diameter). The optical radiant power at the position of the cryogenic radiometer cavity varied with the selection of the light source and optical components in the system. In this case, two 600 groove/mm ruled diffraction gratings blazed at 500 nm were installed in the monochromator system. The LDLS yielded an optical radiant power at the cryogenic radiometer of 2.3 μW at 300 nm and 5.3 μW at 400 nm. When the tungsten lamp was aligned to the monochromator input, the optical radiant power at the cryogenic radiometer was 1.2 μW at 450 nm and 2.9 μW at 1000 nm.

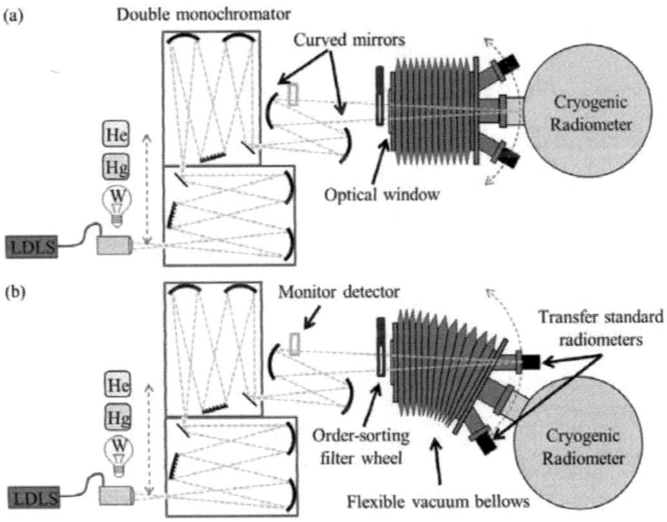

Figure 1. Schematic diagram of the optical power and spectral responsivity measurement apparatus. A motion platform aligns the output light from the monochromator with (a) the cryogenic radiometer or (b) a transfer standard detector. Helium and mercury light sources are used for monochromator wavelength calibration only.

In terms of functionality, this new facility has two main advantages over the previous measurement apparatus: the new cryogenic radiometer and the customized transfer standard mounting platform. The closed-cycle helium cryocooler incorporated into the cryogenic radiometer design gives a greatly increased, uninterrupted run time when compared to traditional liquid helium cooled systems which require refilling of cryogens. The new apparatus has the option of mounting two transfer standard radiometers instead of one, with potential doubling of measurement capability per calibration cycle with the benefit of increased measurement redundancy.

3. Sources of uncertainty due to monochromator system

Primary optical power and spectral responsivity scales can be realized with different sources of optical radiation such as the use of multiple single line lasers and tunable laser sources [2–5]. Several combinations of monochromators and broadband light sources, including Argon arc plasma sources [6], Xe discharge lamps [7], and quartz-tungsten-halogen lamps [1, 8] have been implemented to span a wide spectral range in cryogenic radiometer measurements. When compared to laser-based primary realizations of optical radiant power, a monochromator-based approach offers the versatility of wavelength selection at the expense of laser line wavelength accuracy. As discussed below, some of the systematic differences between these methods arise due to the impact within our method of monochromator wavelength calibration, bandwidth effects, and stray light.

3.1. Wavelength calibration

For the present work, two 600 groove/mm ruled diffraction gratings blazed at 500 nm were installed in the double-subtractive monochromator system. Various spectral lines of helium and mercury lamps were used to calibrate the wavelength readout of each monochromator separately. The spectral lineshapes were fitted to a monochromator slit function, which for our case of a slit input and a round output is given by [9]:

$$I(\lambda) = \frac{1}{2} - \frac{2}{\pi}\left(\frac{2|\lambda-\lambda_0|}{\Delta} - 1\right)\left[\frac{|\lambda-\lambda_0|}{\Delta} - \frac{(\lambda-\lambda_0)^2}{\Delta^2}\right]^{\frac{1}{2}} - \frac{\sin^{-1}[(2|\lambda-\lambda_0|/\Delta)-1]}{\pi} \tag{1}$$

where Δ is the monochromator bandwidth. Polynomial fits to the difference between published spectral line wavelengths and measured wavelengths were used to determine the wavelength correction curves for each of the two diffraction gratings. After the wavelength correction curves were applied to each monochromator, the residual wavelength uncertainty was found to be ± 0.05 nm. The magnitude of the corresponding uncertainties in responsivity were determined using the derivatives of the spectral responsivity curves for the transfer standard radiometers. Absolute rotary encoders mounted on the shaft of each motor are used to repeatably position the gratings (± 0.006 nm).

3.2. Monochromator bandwidth effects

The monochromator slit width was set to 1.0 mm, giving a bandpass of 4.4 nm. The effects of a finite bandwidth in this monochromator-based photodetector calibration system were evaluated and incorporated into the uncertainty budget using a similar approach to that described in Ref. [9]. In the case of a spectrally neutral reference detector, such as the cryogenic radiometer, bandwidth errors can be calculated using the measured spectral responsivity of a detector under test:

$$S_m^t(\lambda_0) = \frac{\int_{\lambda_0+\Delta}^{\lambda_0-\Delta} S^t(\lambda) I(\lambda) P(\lambda) d\lambda}{\int_{\lambda_0+\Delta}^{\lambda_0-\Delta} I(\lambda) P(\lambda) d\lambda} \tag{2}$$

where S^t and S_m^t are the true and measured spectral responsivity of the detector under test, $P(\lambda)$ is the spectral distribution of the monochromator output, and $I(\lambda)$ is the monochromator slit function given in Equation 1. The relative bandwidth error is given by:

$$\epsilon(\lambda_0) = \frac{[S^t_m(\lambda_0) - S^t(\lambda_0)]}{S^t(\lambda_0)}. \quad (3)$$

The unknown quantity S^t is required to solve Equations 2 and 3. A cubic spline interpolation of the measured spectral responsivity data was used in place of S^t, and the corresponding calculated uncertainties due to bandwidth effects from 300 nm to 1000 nm are shown in Table 1.

3.3. Monochromator stray light

Out of band stray light can be caused by scattering, diffraction, or multiple reflections from inside the monochromator. Although stray light levels are usually considered negligible for double monochromators, we decided to test this assumption for the new apparatus. The photocurrent of a single-element silicon transfer standard radiometer was measured from 250 nm to 700 nm with and without an RG665 long pass filter in the beam (Fig. 2). The stray light component of the signal was found to be relatively insignificant: <0.01 % for the LDLS above 300 nm and <0.01 % for the tungsten lamp above 400 nm. The tungsten lamp was only used for spectral responsivity measurements above 400 nm. As the cryogenic radiometer cavity and transfer standard radiometers are coplanar and have similar sized input aperture diameters, the effect of stray light was considered to be negligible.

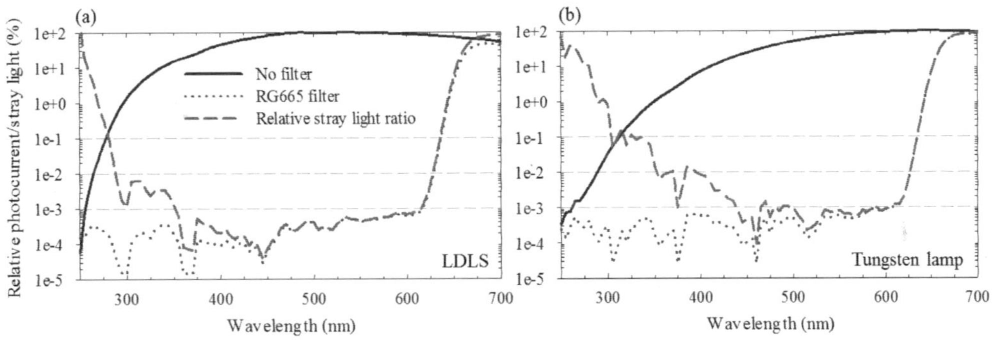

Figure 2. Measured stray light levels in the double monochromator system with (a) the LDLS and (b) the tungsten lamp. A 280 nm long pass filter was mounted to the LDLS output for all measurements to minimize exposure of optical components to UV radiation.

4. Measurement uncertainties

NRC's new absolute radiometry facility has demonstrated significant improvements to its realization of the optical power scale. A complete uncertainty budget for calibrations of single-element silicon transfer standard radiometers using a monochromator and broad-band light source includes wavelength calibration, bandwidth, and temperature effects. These spectrally-dependent uncertainties for spectral responsivity calibrations are listed in Table 1. The repeatability component was estimated by taking the standard deviation of eight measurement cycles. The non-uniformity of the transfer standard detectors has been previously characterized [1] and is accounted for indirectly by the measurement repeatability. The transfer radiometer photocurrent measurement uncertainty component was estimated from the specifications for

IOP Conf. Series: Journal of Physics: Conf. Series **972** (2018) 012014 doi:10.1088/1742-6596/972/1/012014

the digital multimeter (Agilent Technologies 3458A Multimeter) and from the characterization of the NRC-built transimpedance amplifiers used to convert the transfer standard detector photocurrent to a voltage signal.

Table 1. Uncertainty budget for monochromator-based calibrations of single-element silicon transfer standard radiometers (%).

Source of Uncertainty / Wavelength (nm)	Cryogenic Radiometer Effects	Transfer Radiometer Photocurrent Measurement	Measurement Repeatability	Mono. Wavelength Calibration (± 0.05 nm)	Mono. Bandwidth Effects	Transfer Radiometer Temperature (± 0.5°C)	Overall Uncertainty (%)
300	0.011	0.004	0.019	0.013	0.0094	0.008	0.028
320	0.011	0.004	0.008	0.011	0.0096	0.003	0.021
340	0.011	0.004	0.007	0.002	0.0249	0.001	0.029
360	0.011	0.004	0.004	0.003	0.0354	0.016	0.041
380	0.011	0.004	0.006	0.020	0.0180	0.015	0.034
400	0.011	0.004	0.004	0.019	0.0093	0.014	0.028
450	0.011	0.004	0.004	0.009	0.0002	0.009	0.018
500	0.011	0.004	0.004	0.007	0.0006	0.006	0.016
550	0.011	0.004	0.005	0.006	0.0001	0.004	0.015
600	0.011	0.004	0.007	0.005	0.0001	0.004	0.015
650	0.011	0.004	0.007	0.005	0.0000	0.004	0.015
700	0.011	0.004	0.006	0.004	0.0000	0.003	0.014
750	0.011	0.004	0.011	0.004	0.0000	0.004	0.017
800	0.011	0.004	0.014	0.004	0.0000	0.002	0.019
850	0.011	0.004	0.018	0.003	0.0002	0.001	0.022
900	0.011	0.004	0.008	0.004	0.0005	0.002	0.015
950	0.011	0.004	0.010	0.001	0.0027	0.009	0.018
1000	0.011	0.004	0.010	0.006	0.0061	0.128	0.129

Table 2 summarizes the changes in the uncertainty budget between NRC's previous and new absolute cryogenic radiometers. Contributions due to cavity absorptance and non-equivalence effects have had the most significant impact.

Table 2. Uncertainties due to cryogenic radiometer (k=1).

Source of Uncertainty	Magnitude-Old [10] (%)	Magnitude-New (%) [11]
Cryogenic radiometer cavity absorptance	0.01	0.001
Cryogenic radiometer electrical-power measurement	0.01	0.011
Cryogenic radiometer nonequivalence effects	0.01	0.002
Overall Cryogenic Radiometer Effects	0.017	0.011

5. Conclusions

NRC has a new state-of-the-art facility for the primary realization of optical radiant power. A new absolute cryogenic radiometer provides for significant improvements in uncertainty for the realization of the optical power scale, significantly reducing the overall cryogenic radiometer uncertainty contribution from the 0.017 % to 0.011 %. A wavelength dependent uncertainty budget from 300 nm to 1000 nm for silicon detector calibrations was presented. Future work will include the extension of calibration capabilities into the ultraviolet and infrared wavelength ranges with appropriate monochromator diffraction gratings and optical window for the vacuum bellows system.

6. Acknowledgments

The authors would like to acknowledge Jason Demers and the team at NRC Design and Fabrication Services for their contributions to the design, manufacture, and assembly of the motion platform system, transfer detector mounts, and vacuum bellows components.

7. Disclaimer

The commercial equipment identified in this paper are for informative purposes only. Such identification does not imply recommendation or endorsement by NRC, nor does it imply that the equipment is necessarily the best available for the purpose.

References

[1] Boivin L P and Gibb K 1995 *Metrologia* **32** 565
[2] Gentile T R, Houston J M, Hardis J E, Cromer C L and Parr A C 1996 *Appl. Opt.* **35** 1056–1068
[3] Werner L, Fischer J, Johannsen U and Hartmann J 2000 *Metrologia* **37** 279
[4] Eppeldauer G P, Larason T C, Houston J M, Vest R E, Arp U and Yoon H W 2014 *Metrologia* **51** S252
[5] Hong K S, Park S, Kim S K, Park S N and Lee D H 2015 *Journal of the Korean Physical Society* **67** 2045–2058
[6] Meindl P, Klinkmüller A E, Werner L, Johannsen U and Grützmacher K 2006 *Metrologia* **43** S72
[7] Corróns A, Fontecha J L, Corredera P, Campos J, Pons A and Hernanz M L 2000 *Metrologia* **37** 555
[8] Schrama C A, Bosma R, Gibb K, Reijn H and Bloembergen P 1998 *Metrologia* **35** 431
[9] Boivin L P 2002 *Appl. Opt.* **41** 1929–1935
[10] Boivin L P 1998 *Metrologia* **35** 363
[11] L-1 Standards and Technology, Inc 2015 Cryorad-III characterization results

Unidimensional photocurrent model for induced-junction photodiodes

E Borreguero, A Ferrero, J Campos, A Pons and M L Hernanz

Instituto de Óptica (IO, CSIC), Serrano 144, Madrid 28006, Spain

e-mail: elisa.borreguero@csic.es

Abstract. At present, numerical simulations are the common method to predict the response of quantum efficient detectors based on induced-junction diodes. As alternative, this work proposes an analytical model based on photocurrent analysis where surface and bulk losses are modelled considering design and operation photodiode parameters and the characteristics of the incident beam. The model shows how, at short wavelengths, the surface recombination velocity dominates the losses that are directly proportional, whereas the bulk doping and reverse bias voltage determine the losses at long wavelengths. The results obtained by the here proposed analytical model are discussed for different values of surface recombination velocity and compared with simulations reported by other authors. For wavelengths from 400 nm up to 700 nm, the losses calculated by the analytical model at room temperature are in the order of tens of ppm for values between 1000 cm/s and 10000 cm/s of surface recombination velocity. Some simulations agree with a maximum difference of 80 ppm up to 700 nm, where the abrupt rise of bulk losses starts for the analytical model but it is shifted around 800 nm for reported simulations.

1. Introduction

In the last years, within the European joint projects "Candela: Towards quantum based photon standards" (qu-Candela) and "New primary standards and traceability for radiometry" (NEWSTAR), with the development and improvement of self-induced junction photodiodes [1] and the packaging in a wedged trap configuration, a high efficient light-detecting instrument called predictable quantum efficient detector (PQED) [2, 3] has been achieved. The potential of PQEDs as optical radiant power primary standard, based on photoelectric effect in silicon photodiodes, has been proved through comparison to the most accurate primary standard at present, a cryogenic electrical substitution radiometer.

The internal quantum efficiency (IQE) represents the number of electrons contributing to the measured photocurrent versus the number of photons absorbed; therefore, the ideal value of IQE is one assuming a unity quantum yield what would mean absence of internal losses. The IQE of a PQED is predictable from fundamental constants and the radiation wavelength, with small deviations from the ideal behaviour due to internal losses that can be calculated.

Nowadays, simulations performed by software for semiconductor devices are the only proved method to predict the internal losses of a PQED as it is shown by Gran *et al.*, by means of PC1D [4], and Tang *et al.*, by means of Cogenda Genius [5], with a subsequent new implementation of the surface charge. The new implementation of this last reference principally reduces the surface losses by around two orders of magnitude with the same surface recombination velocity.

This work proposes an alternative analytical method, based on the model of the photocurrent analysis developed by Ferrero *et al.* [6], which considers the different internal regions of the photodiode and the characteristics of the incident beam. This work also presents the adaptation of the original model, which considers a physical junction with different doping type on each side, to the self-induced junction of a PQED photodiode, caused by the surface charge resultant of growing a silicon oxide layer on a p-type doped substrate.

2. Internal quantum efficiency and analytical model for induced-junction photodiodes

The produced electrons within every region of the photodiode, dependant on the wavelength of the incident radiation, behave differently, being affected by differentiated surface and bulk losses. In order to predict the IQE of a PQED, it is necessary to quantify the losses due to the radiation not absorbed in the depletion region, where recombination losses are considered to be negligible.

If the whole radiation was absorbed within the depletion region, for each absorbed photon a carrier would contribute to the photocurrent and the IQE would reach its ideal value.

Following this reasoning, IQE of a photodiode can be predicted by the equation (1) quantifying the surface losses (Δ_{surf}) due to the radiation absorbed at the interface oxide-substrate, bulk losses (Δ_{bulk}) due to the radiation absorbed in the bulk beyond the depletion region and losses due to the radiation which goes through the photodiode without being absorbed (Δ_{out}):

$$IQE = 1 - \Delta_{surf} - \Delta_{bulk} - \Delta_{out} \ . \tag{1}$$

In the following subsections the meaning of each item in equation (1) is described and how to adapt them to the self-induced photodiodes used as detectors in PQEDs is explained.

2.1. Internal regions of a self-induced photodiode

For an induced-junction photodiode, schematically represented in figure 1, the parameters of the original model [6] require to be redefined mainly to include the surface charge (Q_{SS}), resultant from growing a silicon oxide layer on a doped p-type substrate, and to eliminate the front recombination region, which is meaningless here.

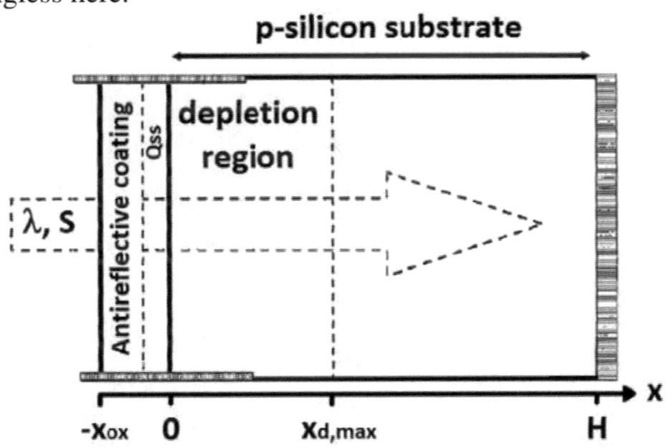

Figure 1. Simplified diagram of a self-induced junction structure in a p-type silicon substrate. The silicon oxide layer, grown on the substrate, induces the surface charge Q_{SS} producing a free charge depletion region.

2.1.1. Surface losses. Bearing in mind the separated contribution of the surface and bulk recombination to the effective lifetime of the charge carriers (τ_{eff}) given by [7]:

$$\frac{1}{\tau_{eff}} = \frac{1}{\tau_{bulk}} + \frac{2S_{eff}}{H}, \tag{2}$$

where τ_{bulk} is the bulk carriers' lifetime, S_{eff} the surface recombination velocity and H the silicon wafer thickness; the losses produced at the interface between the oxide and the silicon can be modelled by equation (3), which is an expression based on a comparison with the saturation current density factors [8, 9] divided by the photocurrent assuming a unity quantum yield:

$$\Delta_{surf} = \left(\frac{n_i^2 S_{eff} H}{Q_{ss}} S \right) \Big/ \left(\frac{n\lambda}{hc} P\,(1-\rho) \right), \tag{3}$$

where n_i is the intrinsic concentration of carriers, S the cross section of the beam, n the refractive index of air, λ the wavelength of the incident radiation in air, h the Planck's constant, c the speed of light in vacuum, P the optical power of the beam and ρ the reflectance of the detector.

2.1.2. Bulk losses. In the bulk, the recombination rate depends on the fraction of the optical radiation absorbed beyond the depletion region (δ_{bulk}) and the probability of finding free recombination centres in this region (ε_{bulk}), as it is described in equation (4). Therefore, ε_{bulk} takes values between 0 and 1 and depends on the bulk dopant concentration (N_A), bulk carriers' lifetime (τ_{bulk}) and the irradiance in the way mentioned in Ferrero *et al.* model [6]:

$$\Delta_{bulk} = \frac{\delta_{bulk}}{2}\left[1 + \varepsilon_{bulk}(\lambda, N_A, \tau_{bulk}, P, S)\right], \tag{4}$$

$$\delta_{bulk} = \exp(-\alpha_{si} x_{d,\max}) - \exp(-\alpha_{si} H), \tag{5}$$

where α_{si} is the absorption coefficient of silicon and $x_{d,max}$ the width of the depletion region. The width of the depletion region of a self-induced photodiode can be modified by changing the reverse bias applied and the bulk dopant concentration [10], according to the equation (6):

$$x_{d,\max\ (bias)} = \left(\frac{2\varepsilon_{r_{si}} \varepsilon_0}{qN_A} \left(V_{s(inv)} - V_{bias}\right) \right)^{1/2}, \tag{6}$$

with q the elementary charge constant, ε_{rsi} the relative permittivity of silicon, ε_0 the vacuum permittivity, V_{bias} the bias voltage applied through the photodiode and $V_{s(inv)}$ the strong inversion surface potential, considered equal to the double of the bulk Fermi potential [10]:

$$V_{s(inv)} = 2\frac{k_B T}{q} \ln \frac{N_A}{n_i}, \tag{7}$$

with Botlzmann's constant, k_B, and temperature, T.

2.1.3. Not absorbed radiation losses. Finally, the radiation which goes through the photodiode without being absorbed is given by:

$$\Delta_{out} = \exp(-\alpha_{si} H). \tag{8}$$

Reflection from the back surface is not taken into account in this model because the radiation which goes through the photodiode without being absorbed, without considering back reflection, is negligible between 400 nm to 900 nm compared with surface and bulk losses.

3. Results of the analytical model

3.1. Input data

The aim of this work is to propose an analytical expression for equation (1) as an alternative to software simulations in order to determine the internal losses of PQEDs. As seen above, the expressions of the analytical model depend on design and operation parameters of the device whose values in the case of photodiodes used in PQEDs are shown in table 1, taken from reference [2], so that result yields by this model can be compared to those published from the numerical model [4, 5].

Table 1. Reference design and operation parameters of the induced-junction photodiodes used in PQEDs.

Parameter	Symbol	Reference value
Wafer thickness	H	525 µm
Surface charge density	Q_{ss}	6.2×10^{11} cm^{-2}
Bulk dopant concentration	N_A	2×10^{12} cm^{-3}
Reverse bias voltage	V_R	-5 V
Optical power	P	100 µW
Beam cross section	S	10 mm^2

3.2. Spectral behaviour

As PQEDs are highly efficient detectors, it is more interesting to present the results of the internal quantum deficiency (IQD), defined as the default to unit. The following graphs represent the spectral IQD, in parts per million (ppm), predicted by the analytical model for a single photodiode with design parameters from table 1. In order to verify whether the model yields expected results, the effects of reverse bias voltage, surface recombination velocity, bulk dopant concentrations and bulk carriers' lifetime over IQD have been calculated.

3.2.1 Reverse bias voltage and surface recombination velocity influence. Using parameters in table 1, with different values of reverse bias voltage and surface recombination velocity, the results shown in figure 2 are obtained for IQD. Increasing the reverse bias voltage extends the depletion region depth and, hence, shifts the rise of the bulk losses towards longer wavelengths; whereas, increasing the surface recombination velocity by one order of magnitude produces an increase of the same order of the surface losses, this effect is illustrated in figure 2 by the represented IQD at short wavelengths.
This behaviour is similar to that it is shown by Gran *et al.* [4] and Tang *et al.* [5] in their reported simulations. PC1D simulations [4] show an IQD curve, calculated for a 10 cm/s value of surface recombination velocity, very close to the 1000 cm/s curve in figure 2. On the other hand, Cogenda Genius TCAD simulations [5], which have a surface recombination two order higher than PC1D simulations for the same IQD results, show an IQD from 80 ppm (at 400 nm) up to 1 ppm (at 800 nm) calculated for a 4500 cm/s value of surface recombination velocity. Comparing with the analytical model, this Cogenda Genius TCAD IQD curve is included between the 1000 cm/s and 10000 cm/s curves in figure 2. Around 700 nm, the main difference is found when the starting point of the abrupt rise of the bulk losses appears for the analytical model but for the Cogenda Genius TCAD simulations is shifted at longer wavelengths around 800 nm, in spite of using both the same depth of the depletion region in absence of radiation. Therefore, it can be extracted that, the IQD results predicted by the analytical model agree with Cogenda Genius TCAD simulations with a maximum difference of 80 ppm, between 400 nm and 700 nm.

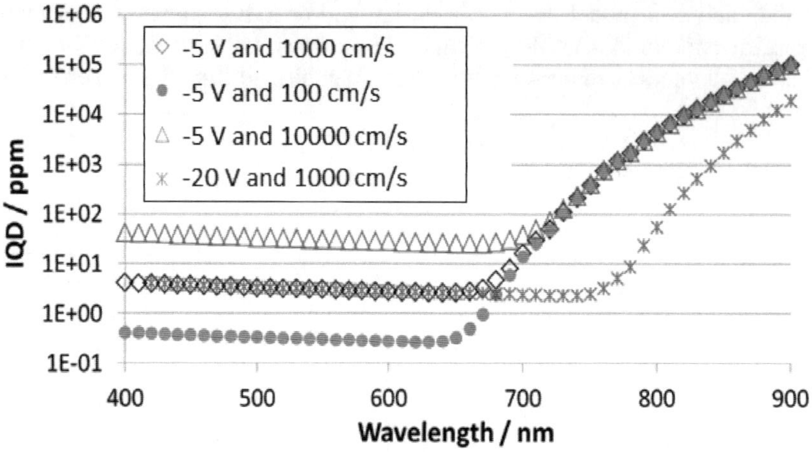

Figure 2. Four scenarios of spectral results of IQD calculated at 300 K by the analytical model with parameters in table 1 using 0.4 ms of bulk carriers' lifetime and different reverse bias voltages and surface recombination velocities.

3.2.2 Bulk dopant concentration and bulk carriers' lifetime influence. Following the same strategy, but changing bulk dopant concentration and the bulk carriers' lifetime rather than reverse bias voltage and surface recombination velocity, IQD values shown in figure 3 can be observed. The increase of the bulk doping shortens the depletion region depth and shifts the rise of the bulk losses towards shorter wavelengths; whereas, increasing the bulk carriers' lifetime reduces the slope of the bulk losses. The reason is that dopant impurities behave as recombination centres increasing the bulk losses.
The influence of the bulk dopant concentration and the shape of the IQD curves are similar to the reported simulations [4, 5], again with the exception of the rise of bulk losses, which starts at shorter wavelength for the analytical model.

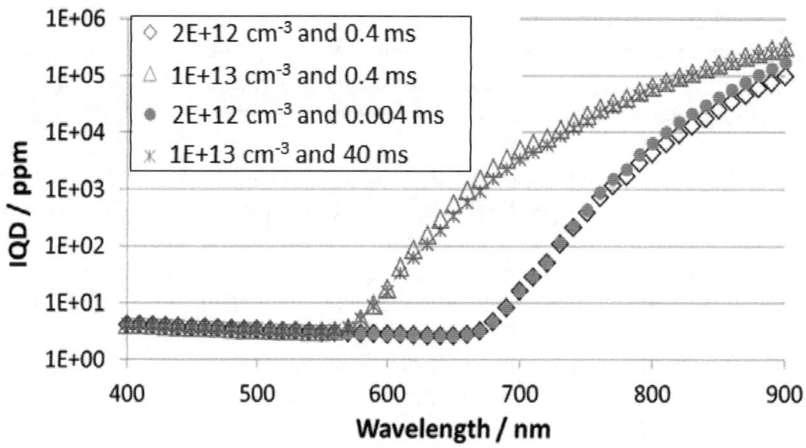

Figure 3. Four scenarios of spectral results of IQD calculated at 300 K by the analytical model with parameters in table 1 using 1000 m/s of surface recombination velocity and different bulk dopant concentrations and bulk carriers' lifetimes.

4. Conclusions

An analytical model to predict the IQE of a PQED single induced-junction photodiode has been developed by adapting the Ferrero *et al.* model [6]. This analytical model is useful to quantitatively determine how the design and operating parameters of the photodiodes contribute to the IQE. This work proposes the analytical model as an alternative method to predict the IQE of a PQED radiometer, since its behaviour is similar to the software simulations.

Qualitatively, the shape and the behaviour of the IQD curve predicted by the analytical model agree with the simulations performed by Gran *et al.* by means of PC1D [4] and Tang *et al.* by means of Cogenda Genius TCAD [5]. At short wavelengths, the surface recombination velocity dominates the losses being directly proportional, then, as the radiation penetrates into the depletion region, the losses drop slightly and, finally, the IQD increases quickly at longer wavelengths due to the abrupt rise of the bulk losses determined mainly by the bulk doping and reverse bias voltage.

Quantitatively, the IQD predicted by the analytical model is in the order of tens of ppm for values between 1000 cm/s and 10000 cm/s of surface recombination velocity, for wavelengths from 400 nm up to 700 nm. Comparing these results with simulations, the curves of the IQD predicted by the analytical model and Cogenda Genius TCAD simulations, which reduce the surface losses by around two orders of magnitude with the same surface recombination velocity respect PC1D simulations, both agree with a maximum difference of 80 ppm for wavelengths between 400 nm and 700 nm. However, around 700 nm, the main difference appears when the rise of bulk losses starts for the analytical model but it is shifted at longer wavelengths around 800 nm for the reported simulations, in spite of using both the same depth of the depletion region in absence of radiation.

For future improvements, it is required to study deeper the mechanisms of the minority carriers diffusion and the influence of the irradiance over the width of the depletion region.

References

[1] Hansen T E 1978 *Physica Scripta* **18** 471-5
[2] Sildoja M, Manoocheri F, Merimaa M, Ikonen E, Müller I, Werner L, Gran J, Kübarsepp T, Smîd M and Rastello M L 2013 *Metrologia* **50** 385-94
[3] Müller I *et al* 2013 *Metrologia* **50** 395-401
[4] Gran J, Kübarsepp T, Sildoja M, Manoocheri F, Ikonen E and Müller I 2012 *Metrologia* **49** 130-4
[5] Tang C K, Nordsveen M U and Gran J 2017 *Proc. of the 13th Int. Conf. on New Developments and Applications in Optical Radiometry NEWRAD 2017* (Tokyo) pp 142-143
[6] Ferrero A, Campos J, Pons A and Corrons A 2005 *Applied Optics* **44** 208-16
[7] Kerr M J and Cuevas A 2002 *Semiconductor Sci. Technol.* **17** 35-8
[8] Kimmerle A, Rothhardt P, Wolf A and Sinton R A 2014 *Energy Procedia* **55** 101-6
[9] Kane D E and Swanson R M 1985 *Proc. of the 18th IEEE Photovoltaic Specialists Conf.* (Las Vegas) pp 578-583
[10] Grove A S 1967 *Physics and Technology of Semiconductor Devices* (New York: Wiley) pp 157-161

Acknowledgments

This work was supported by the Comunidad de Madrid and the European Union under program S2013/MIT-2790-SINFOTON and by the European Commission under EMRP JRP SIB57-NEWSTAR. The EMRP is jointly funded by the EMRP participating countries within EURAMET and the European Union.

Improved calibration strategy for luminous intensity

P Schneider[1], K Salffner[1], A Sperling[1], S Nevas[1], I Kröger[1], T Reiners[2]

[1]Physikalisch-Technische Bundesanstalt Braunschweig und Berlin, Bundesallee 100, 38116 Braunschweig, Germany
[2]LMT Lichtmesstechnik GmbH Berlin, Helmholtzstrasse 9, 10587 Berlin, Germany

E-mail: philipp.schneider@ptb.de

Abstract. To improve the realization of luminous intensity a new photometric detector has been developed at Physikalisch-Technische Bundesantalt (PTB). This detector will be used to realise a radiometric traceability chain for luminous intensity with a lowered measurement uncertainty. To allow a short and direct calibration hierarchy, a trap detector was chosen that will be used with several modular attachments for each calibration step. Achieving a reduced measurement uncertainty requires careful design and characterization of each module used. The new calibration strategy and the detector with its several setups are described here.

1. Introduction

The realisation and maintenance of the unit of luminous intensity at Physikalisch-Technische Bundesanstalt (PTB) is done in two complementary ways to provide confidence in the disseminated unit. These two ways are a detector-based radiometric approach and a lamp-based approach. The radiometric realisation of the candela at PTB [1] is based on the following equation for the luminous intensity $I_v(T)$ at the distribution temperature T of an incandescent lamp:

$$I_v(T) = \frac{K_m}{s_0(\lambda_d)} \cdot \frac{r^2}{\Omega_0} \cdot y \cdot F(T) \quad (1)$$

with the spectral-mismatch factor

$$F(T) = \int_0^\infty S(\lambda, T)V(\lambda)\mathrm{d}\lambda \cdot \left(\int_0^\infty S(\lambda, T)s_r(\lambda)\mathrm{d}\lambda\right)^{-1} \quad (2)$$

Here $K_m = 683$ lm/W is the luminous efficacy of radiation at 540 THz, $s_0(\lambda_d)$ is the absolute irradiance responsivity of the photometer at wavelength $\lambda_d = 555$ nm, r is the distance between the incandescent lamp and the photometer, y is the measured photocurrent, $V(\lambda)$ is the luminous efficiency function as defined by CIE, $s_r(\lambda)$ is the relative spectral responsivity of the photometer normalized at λ_d and $S(\lambda, T)$ is the spectral distribution of the incandescent lamp at distribution temperature T.

The two complementary ways to realize the unit of luminous intensity was already described by Erb and Sauter [2]. The radiometric realization of the unit is traceable to the cryogenic radiometer via several steps which will be presented in more detail in this paper as the base of the new improved realization strategy. This radiometric realization of the unit results in an expanded uncertainty for the luminous intensity of $U(I_v) = 2.5 \times 10^{-3} (k = 2)$. In addition, the unit is maintained by a network of lamps. This network still consists of those lamps which were involved during the comparison of realizations of the candela in the late 1970th which finally led to the defined constant in the current candela definition.

Mainly the superior reproducibility of these lamps determines the uncertainty of the maintained unit at PTB. With an additional network of photometers, the maintained luminous intensity is currently provided at an uncertainty level of $U(I_\mathrm{v}) = 2 \times 10^{-3} (k = 2)$. When comparing both ways to establish the unit it becomes apparent that despite the comparable uncertainties the reproducibility of the radiometric realization is worse than the reproducibility of the maintained unit as described elsewhere in more detail [3]. This implies that using the lamp based route the uncertainty of the calibration of a luminous intensity standard can be even smaller while such a calibration using the detector based route would be even larger. This discrepancy poses the requirement of an improved radiometric calibration strategy with reduced measurement uncertainty, thus ensuring a more stable detector based realization of the unit of luminous intensity in future.

2. Steps of the radiometric realization

Multiple steps are necessary to determine the aforementioned radiometric properties of a photometer. With known absolute irradiance responsivities $s(\lambda)$ the photometric responsivity with respect to CIE Standard Illuminant A based on a distribution temperature of $T_\mathrm{A} = 2856$ K can be determined following the formula:

$$s_\mathrm{v} = \frac{1}{K_\mathrm{m}} \int_0^\infty S(\lambda, T_\mathrm{A}) s(\lambda) \mathrm{d}\lambda \cdot \left(\int_0^\infty S(\lambda, T_\mathrm{A}) V(\lambda) \mathrm{d}\lambda \right)^{-1} \quad (3)$$

The calibration hierarchy for the photometric responsivity of a photometer using the current realization is shown in comparison to the new calibration strategy in figure 1.

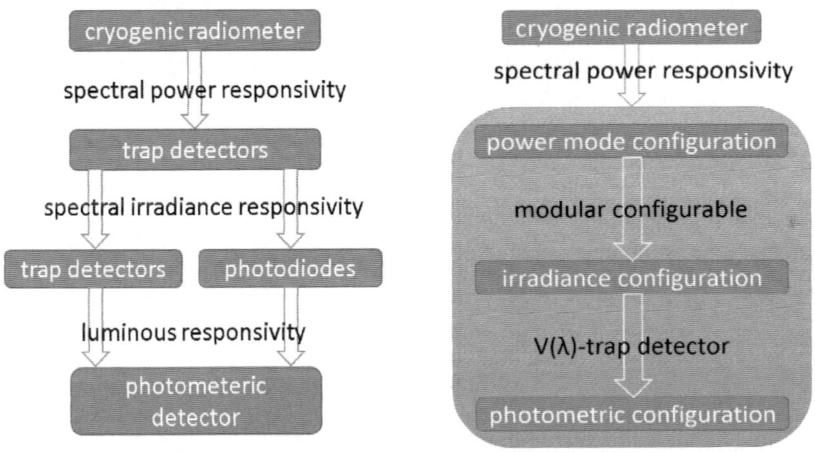

Figure 1. Radiometric calibration hierarchy for photometric detectors. (left): The current calibration hierarchy requiring multiple transfer detectors to achieve a traceable luminous responsivity of a photometer. (right): The new shorter calibration hierarchy utilising the different setups of a new modular detector. All steps highlighted in the blue box are done with the same detector, decreasing the uncertainty.

The first step in the traceability chain is the calibration of silicon trap detectors with a cryogenic radiometer regarding their spectral power responsivity as described by Werner et al. [4]. The spectral power responsivity is measured at several cw laser wavelengths across the visible spectral range and interpolated by the model mentioned in [4]. The spectral power responsivity is transferred to additional trap detectors using spectrally tunable measurement setups with lamp-monochromator based light sources. By adding mechanically manufactured precision apertures the spectral irradiance responsivity of the trap detectors and photodiodes can be determined. The calibrated trap detectors and photodiodes are used as reference detectors at two setups in PTB. The DSR setup [5] and the TULIP setup [6] can both be used to calibrate the spectral irradiance responsivity of photometric detectors. The DSR setup is also operated with a lamp-monochromator based light source and uses photodiodes as reference while the TULIP setup performs measurements with specially treated tunable quasi-cw laser radiation to illuminate the detectors while using the trap detectors as reference detectors.

At both setups, the major contributions to the measurement uncertainty when calibrating a photometer are the uncertainty of the responsivity of the reference trap detector and photodiode, respectively, and the area of the mechanical aperture to enable measurements in irradiance mode. With the monochromator based DSR setup the uncertainty contributions of wavelength and of the spectral bandwidth dominate the combined uncertainty when not measuring close to the peak of the $V(\lambda)$-function. For the TULIP setup, the combined measurement uncertainty is mainly influenced by the aforementioned uncertainty of the responsivity of the reference and the aperture as well as by the uncertainty of the wavelength determination.

3. Modular photometric $V(\lambda)$-trap detector system

To enable the usage of a shorter traceability route with reduced measurement uncertainty a new modular detector system was developed. The detector can be setup in optimized configurations for each of the steps shown in the blue box in figure 1. However, avoiding the additional transfer steps of the current traceability chain already offers a reduced uncertainty for the photometric responsivity. The core of the $V(\lambda)$-trap detector is a six-diode transmission trap detector similar to the one described by Kübarsepp et al. [7]. For the photometric trap detector six photodiodes with 18×18 mm² active area were used enabling a slightly larger acceptance angle compared to the 10×10 mm² photodiode setup from Kübarsepp when used with the same aperture of 6 mm. The maximum usable active area perpendicular to the optical axis is approximately 150 mm². It is defined by the size and positioning of the six photodiodes in the trap detector. A light cone entering the detector aperture must not exceed this area due to divergence so that all light is only incident on the diodes. When using the 6 mm aperture the optical path length within the trap of 160 mm from the aperture to the last diode results in a half acceptance angle of only 1.2 degree.

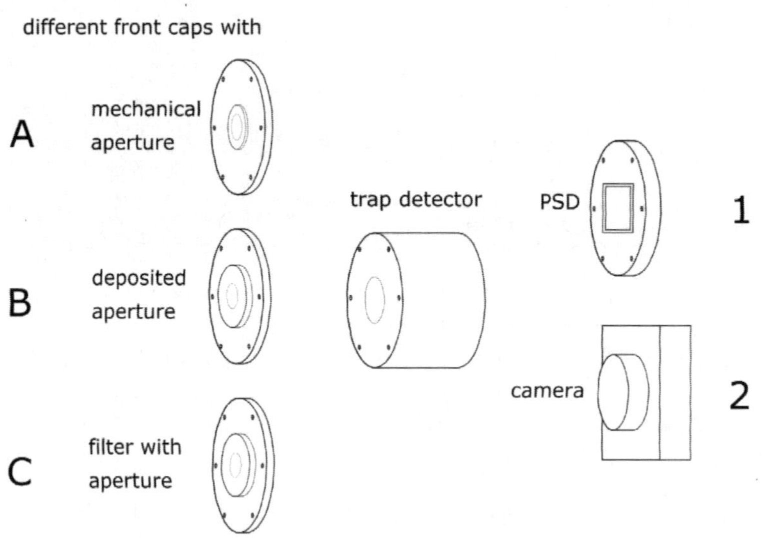

Figure 2. Assembly drawing of the six-element transmission trap detector with front caps A, B and C with different aperture and filter elements and the rear modules 1 and 2 that can be moved into the transmitted beam.

IOP Conf. Series: Journal of Physics: Conf. Series **972** (2018) 012016 doi:10.1088/1742-6596/972/1/012016

The six-element transmission trap detector is in theory independent of polarization effects and, thus, suitable for polarized laser beam as well as for incandescent lamp measurements. To carry out each of the steps in the calibration strategy, different attachments can be mounted to the front side of the trap detector in the optical pathway of the incoming beam or to the rear side of the detector to measure the transmitted beam. The attachable modules and their use during calibration are described in the following sections and depicted in figure 3.

3.1. Front modules of the detector

A front cap with a mechanically manufactured aperture (see figure 2, module A) will be attached to the trap detector for the calibration in radiant power mode with the cryogenic radiometer. The aperture will have the same nominal area as the entrance aperture of the cryogenic cavity. It is not required to have the most precise knowledge of the area of the aperture when measuring the spectral power of incident laser beams. This aperture only affects the outer part of the beam, namely the straylight in forward direction. As the straylight makes only a minor part of the incident power, the usage of nominally equal sized apertures in front of the trap and the cavity is sufficient to match the straylight fractions entering both detectors.

For using the detector in irradiance mode, a precision chrome aperture deposited on a glass substrate (see figure 2, module B) is used. Deposited apertures pose the advantage of having sharper aperture edges and thus the aperture area can be determined with lower uncertainty compared to conventional mechanically manufactured apertures. To enable the transfer from power responsivity to irradiance responsivity with a low uncertainty contribution, the power responsivity of the trap with attached deposited aperture must be determined. The transmittance of the used substrate will be determined in power mode for this purpose via direct comparisons with measurements using module A. As the calibration with the cryogenic radiometer is carried out using cw-laser radiation, effects of the deposited aperture such as transmission and diffraction should be also measured with cw-lasers. For this purpose, the aperture was specially manufactured to have low interference effects. This is achieved by adding a glass wedge to the aperture substrate by means of optical contact bonding. The currently remaining interference effect on the transmission is in the range of a few parts in 10^3 and will be corrected for when using cw-lasers.

To match the responsivity of the detector to the $V(\lambda)$-function, a suitable $V(\lambda)$-filter having a deposited aperture (see figure 2, module C) is used. The area of the aperture is nominally the same for the modules B and C. The transmittance of the filter of module C will be determined by relative measurement with respect to module B with tunable lasers in irradiance mode. The filter and also the deposited aperture were assembled on top of a glass wedge to reduce interference caused by interreflection between the outer interfaces. By using the same angle of 0.5° for the glass wedge in both modules, the diffraction caused by the tilted interface is the same in both cases. The transmittance of such a filter is expected to show a spectrally dependent temperature dependence [8]. To avoid effects of the temperature the filter and the trap are temperature controlled to about 30 °C with a stability of 0.05 K.

3.2. Rear modules of the detector

For the detection (and investigation of the properties) of the light that is transmitted through the trap detector, two different modules can be placed at the rear of the detector.

A position sensitive diode (PSD) (see figure 2, module 1) is used for tracking the position of the beam across the active area during the measurements in optical power mode. This sort of photodiode utilises a PIN diode with a special electrode configuration. The position of the centroid of an incident light beam can be tracked by the change of the ratios of the photocurrents at the electrodes. The PSD will be used during the calibration of the trap detector with the cryogenic radiometer and the transmittance measurement of the aperture substrate. This way, the position of the beam can be tracked, which allows to account for the nonuniformity of the detector when switching from power to irradiance mode yielding the irradiance responsivity with a lower uncertainty. The nonuniformity of the $V(\lambda)$-trap detector was determined in a separate measurement [3].

In addition, a camera (see figure 2, module 2) with a large CCD of 27.65 mm × 18.48 mm can be used. When using the detector in irradiance mode the transmitted light is detected by the camera. The camera is used for a qualitive observation of the light field and to evaluate the illumination conditions inside the transmission trap, to recognize light hitting edges of the photodiodes. This way it can be assured that the detector is used only in the well-defined irradiance mode where all the light transmitted through the aperture area is seen by all six diodes. Since the geometry of the trap and the aperture allows only small acceptance angles, the camera offers a way to check for appropriate illumination.

4. Characterization of the *V(λ)*-trap detector

In order to show the suitability of the new detector system to reach the aimed uncertainty, several measurements were done to characterize the whole detector system and the trap detector individually before the new calibration strategy will be implemented. To use the detector both, in power and irradiance mode, the uniformity of the spatial responsivity of the detector is of highest importance and it was shown to be in the range of 10^{-4} [3].

As mentioned before, the design of the trap detector yields a theoretically polarization independent responsivity. To check for remaining polarization dependencies, the detector was rotated around its optical axis with a strictly linearly polarized incident laser beam. The relative maximum deviation of the measured photocurrent when rotating the detector by more than 180 degree was in the range of $\pm 4 \times 10^{-5}$. Within the standard deviation of 2×10^{-6}, it could not be distinguished between remaining effects of the mechanical fixation of the trap detector and the polarization dependency.

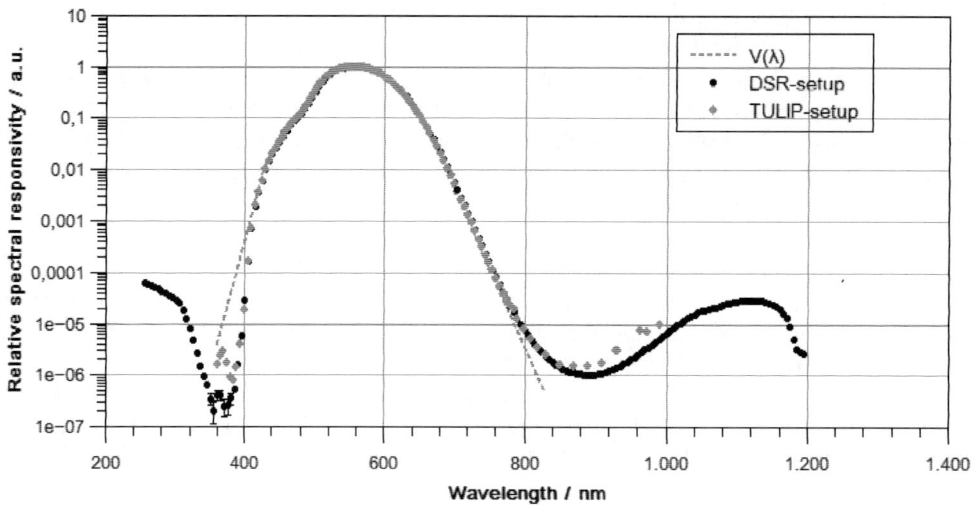

Figure 3. Relative spectral responsivity of the new detector normalized to the values at 555 nm. Calibration results of the irradiance responsivity from the DSR setup are shown in black circles, first results from the TULIP setup in orange squares and the *V(λ)*-function as a dashed green line, respectively. Error bars are shown for the DSR results representing the measurement uncertainties. For TULIP no error bars are visible as they are completely overlapped by the symbols representing the values.

As a preliminary investigation of the spectral properties of the new detector its spectral irradiance responsivity with the *V(λ)*-filter was calibrated at two calibration setups using the current traceability route. The lamp-monochromator based DSR setup was used for calibration in the spectral range of 250 nm to 1200 nm while the measurements at the laser-based TULIP setup were recorded in the spectral range of 360 nm to 1000 nm. To compare the two calibrations, the results of the spectral irradiance responsivity are shown in figure 3. The uncertainty evaluation of the measurement with TULIP is preliminary and requires further investigation of the differences in the UV and NIR spectral range. It is apparent that although the *V(λ)*-function is limited to the wavelength range 360 nm to 830 nm there is a residual responsivity in the infrared and UV spectral range. When using the detector in combination with an incandescent lamp at 2856 K distribution temperature, especially the infrared responsivity of the detector results in a contribution to the total photocurrent at the level of 10^{-4}. As the target uncertainty of the new calibration strategy is within the same order of magnitude, an appropriate correction must be carefully determined and applied. To express the quality of the detector's match to the *V(λ)*-function the f_1' quality factor can be calculated. The calculation according to CIE S023 [9] results to be f_1'= (1.78 ± 0.21) %.

5. Conclusion

A new detector system was set up at PTB allowing for a more direct calibration hierarchy to trace luminous intensity to optical power. The detector system is a modular one. The modules have been chosen carefully to fulfil the requirements of each step of the new calibration strategy. Modules of the *V(λ)*-trap detector were characterized to evaluate their suitability for the improved traceability chain allowing a reduced overall measurement uncertainty for the realization of the luminous intensity unit. Aperture and filter modules were designed and built in order to reduce interference modulation of the transmitted radiation, which is important for the detector calibration at the laser-based TULIP setup. Spectral irradiance responsivity calibrations following the current calibration strategy have shown a good match to the *V(λ)*-function with $f_1' \approx 1.8$ %. The calibration against the cryogenic radiometer following the presented calibration strategy is planned as a next step. The results including the final measurement uncertainty will be reported on in another publication.

Acknowledgements
The author gratefully acknowledges support by the Braunschweig International Graduate School of Metrology B-IGSM.

References

[1] Ohno Y, Goodman T, Sauter G 1999 Trilateral Intercomparison of Photometric Units Maintained at NIST (USA), NPL (UK) and PTB (Germany) *J. of Research of the National Institute of Standards and Technology* **104** 47-57
[2] Erb W and Sauter G 1997 PTB network for realization and maintenance of the candela *Metrologia* **34** 115-24
[3] Schneider P, Sperling A, Nevas S, Salffner K 2015 Developing a new detector for the realization of the SI base unit "candela" *Proc. of CIE Expert Symp. on the CIE S 025 LED Lamps, LED Luminaires and LED Modules Test Standard (Braunschweig)* (Vienna: CIE) 62-70
[4] Werner L, Fischer J, Johannsen U, Hartmann J 2000 Accurate determination of the spectral responsivity of silicon trap detectors between 238 nm and 1015 nm using a laser-based cryogenic radiometer *Metrologia* **37** 279-84
[5] Winter S, Sperling A 2006 Uncertainty analysis of a photometer calibration at the DSR setup of the PTB *Proc. of the 2nd CIE Expert Symp. on Meas. Unc. (Braunschweig)* (Vienna: CIE) 139-42
[6] Schuster M, Nevas S, Sperling A 2014 Validation of short-pulse-laser-based measurement setup for absolute spectral irradiance responsivity calibration *Applied Optics* **53** 13 2815-21
[7] Kübarsepp T, Kärhä P, Ikonen E 1997 Characterization of a polarization-independent transmission trap detector *Applied Optics* **36** 13 2807-12
[8] Kärhä P, Lassila, A, Ludvigsen H, Manoochehri F, Fagerlund H, Ikonen E 1995 Optical power and transmittance measurements and their use in detector-based realization of the luminous intensity scale *Optical Engineering* **34** 9 2611-18
[9] Blattner P, et al. 2013 *Characterization of the Performance of Illuminance Meters and Luminance Meters* CIE S023/E:2013 (Vienna: CIE)

IOP Conf. Series: Journal of Physics: Conf. Series **972** (2018) 012017 doi:10.1088/1742-6596/972/1/012017

Accurate Measurement of New Type Non-silicon Solar Cells' Photoelectric Conversion Efficiency

Haifeng Meng, Limin Xiong, Junchao Zhang, Yingwei He, Bifeng Zhang and Chuan Cai

National Institute of Metrology, Beijing, P. R. China, No.18, Bei San Huan Dong Lu, Chaoyang District, Beijing, P.R.China

menghf@nim.ac.cn

Abstract. Different from traditional silicon solar cells, new type non-silicon solar cells are composed of various kinds of materials, such as organic, perovskite, quantum dots and so on. They always show special performance in spectral responsivity, compatibility and stability, which brought great challenges for the accurate measurement of photoelectric conversion efficiency. We will illustrate a method for accurate efficiency measurement mainly in the view of metrology. Based on effective traceability and calibration of the standard value, a procedure for new type non-silicon solar cells' efficiency measurement would be proposed. Main influencing factors such as spectral mismatch will also be analysed. Hope to provide a reference method for the accurate characterization of the photo-electric properties of various new types of solar cells.

1. Introduction

New type non-silicon solar cells, such as organic, perovskite, quantum dots solar cells, etc., have become hot spots in science research. They are low cost and flexible, and most importantly their photoelectric conversion efficiency is surging up [1-6]. However, they also have unique properties in spectral responsivity, stability and compatibility, which largely make it a challenging task for accurate measurement of efficiency. Lack of accuracy, reliability and mutual recognition for the efficiency data has seriously influenced solar cell research. Mischaracterized and invalid efficiency results have been reported in a number of journals, improvement of measurement accuracy was urgently needed [7-9]. Nature publishing group have announced a reporting checklist for solar cell manuscripts, which composed of ten items like a questionnaire to be answered.

The measurement inaccuracy is largely due to ineffective traceability of standard value and special properties of new type solar cells [10]. For example, a PTB7-Th/PC_{71}BM-based polymer solar cell, when measured under the same solar simulator, but with two different reference solar cells, KG5 filtered or non-filter reference solar cells, the J-V curves are significantly different as shown in Figure 1. The relative deviation of efficiency is 9.3%, which cannot be ignored [11]. Scientists call for an accurate method of efficiency measurement for new type solar cells.

In this paper, combining the effective traceability of standard value with samples' own special properties, we presented an accurate method of current-voltage characteristics measurement for new type non-silicon solar cells. Measurement details, calculation of spectral mismatch factors, and main influencing factors were also illustrated. A whole procedure was proposed for accurate measurement of new type non-silicon solar cells' photoelectric conversion efficiency. It can traceable to SI units, and the route is applicable to most of new type solar cells' efficiency measurement.

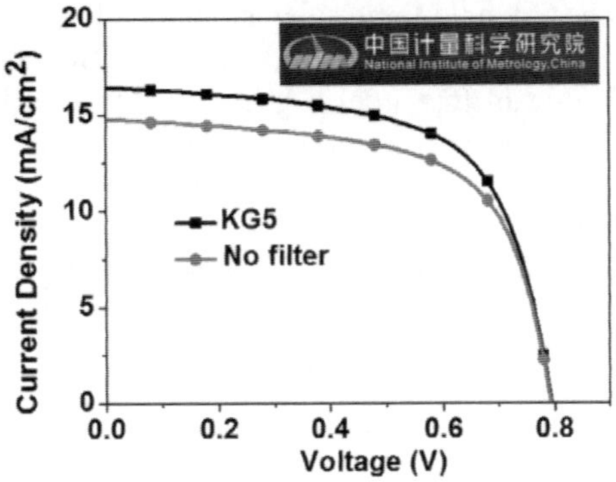

Figure 1. The measured J–V results of a PTB7-Th/PC_{71}BM-based polymer solar cell, when use different reference solar cells to calibrate the same solar simulator [11].

2. Measurement

2.1. Traceability

Photoelectric conversion efficiency is defined as the ratio of output power to input power, it can be calculated out according to the solar cell's *I-V* characteristics under standard test conditions (STC: 1000W/m^2, AM1.5G, 25℃) [12]. Practically, *I-V* characteristics were measured under solar simulators which calibrated to STC state by reference solar cell [13]. As illustrated in Figure 2, reliable calibrated value (CV) and effective traceability to it are critical to efficiency measurement.

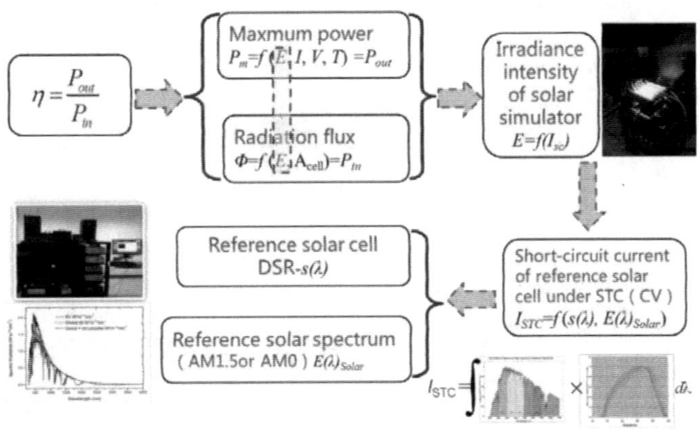

Figure 2. Route map of solar cell's efficiency determination under STC.

The route in Figure 2 is mature for silicon solar cells. While for new type non-silicon solar cells, problems emerged because of spectral responsivity mismatch and other relevant mechanisms. Considering new type non-silicon solar cells' own properties, we developed a measurement procedure as shown in Figure 3. Firstly, measure the spectral responsivity (SR) or external quantum efficiency (EQE). Secondly, choose calibrated reference solar cell and solar simulator simultaneously according to spectral mismatch factor (MMF) data [14], which is a result of mutual matching and coordination. The reference solar cell should be previously calibrate to obtain its CV (I_{sc}-STC) by DSR method [15, 16], which is traceable to SI units through standard detectors. Cryogenic radiometer is the highest standard for this traceability chain. International comparisons of standard detectors and reference solar cells were carried out regularly to achieve international mutual recognition. The solar simulator should be finely adjusted and calibrated according to the requirements of IEC 60904-9, which includes spectral mismatch, non-uniformity and instability of irradiance intensity [17, 18]. During our measurement, a highly matched double-light solar simulator and silicon reference solar cell were chosen because of their excellent MMF data. The practical 1000W/m^2 irradiance and AM1.5G spectra of STC conditions was calibrated in this step to guarantee the effective value traceability [12, 13]. Finally, whole *I-V* curves with critical parameters such as short circuit current, open circuit voltage, and maximum power data, would be obtained after scan parameters set by the measurement software. After the effective area calibrated, photoelectric efficiency would be calculated out. The value is traceable to SI unit, so that guarantee the reliability and mutual recognition.

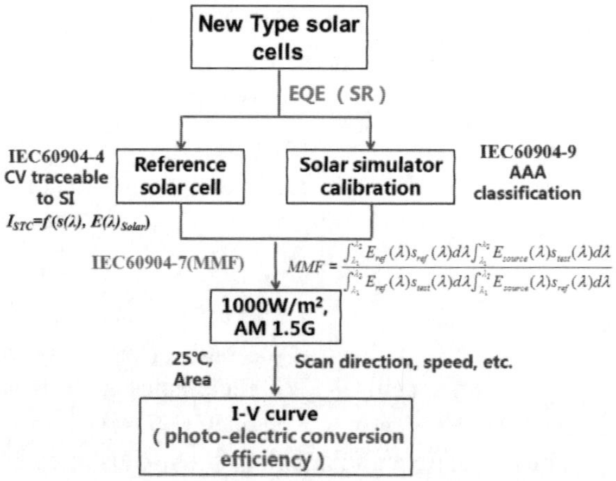

Figure 3. Schematic map for accurate measurement of new type solar cell's efficiency.

2.2. Spectral mismatch factor calculation

Spectral mismatch can be calculated according to equation (1) [14]. If the spectrum of solar simulator $E_{source}(\lambda)$ is the same as reference spectral irradiance distribution $E_{ref}(\lambda)$, or the spectral response of the reference solar cell $s_{ref}(\lambda)$ is the same as the test solar cell $s_{test}(\lambda)$, MMF=1, no modification is needed.

$$\mathrm{MMF} = \frac{\int_{\lambda_1}^{\lambda_2} E_{ref}(\lambda)s_{ref}(\lambda)d\lambda \int_{\lambda_1}^{\lambda_2} E_{source}(\lambda)s_{test}(\lambda)d\lambda}{\int_{\lambda_1}^{\lambda_2} E_{ref}(\lambda)s_{test}(\lambda)d\lambda \int_{\lambda_1}^{\lambda_2} E_{source}(\lambda)s_{ref}(\lambda)d\lambda} \quad (1)$$

For silicon-based solar cells, no matter mono-Si or poly-Si sample, have similar spectral responsivity. Under single-light or double-light solar simulator, the mismatch errors (MMF-1) are all small enough, as results shown in Table 1. We can consider it as an uncertainty component.

Table 1. (MMF-1) calculation results of the mono-Si and poly-Si solar cell measurement, by using the same mono-Si reference solar cells under NIM's two different solar simulators.

Sample	Double-light solar simulator	Single-light solar simulator
Mono-Si	-0.06%	-0.16%
Poly- Si	0.24%	-0.59%

While most of new type non-silicon solar cells, such as organic, DSSC and perovskite solar cells, their spectral responsivity are largely different from silicon reference, leading to non-negligible mismatch error. Take perovskite solar cell as a typical example. Figure 4a demonstrated the spectral responsivity curves of mono-Si reference solar cell (Black) and the sample (Blue), while Figure 4b corresponding to KG5-filter Si reference solar cell (Black) and the same sample (Blue). We can see neither mono-Si nor KG5-filter Si reference solar cell can well-match to its spectral responsivity. Then the spectral distribution of light source became critical for the total MMF calculation. Figure 4c displayed the spectral irradiance of NIM's double-light (Black) and single-light (Red) solar simulator compare to AM1.5G standard spectral irradiance (Green). The double-light solar simulator, which composed of Xeon lamp and halogen lamp with a tailor-made filter, its mismatch with AM1.5G or AM0 can be adjusted to smaller than 5% with 50 nm integral, in the range of 300 nm to 1800 nm. The critical tailor-made filter has two sides which play different roles. One side can make the Xenon lamp irradiance above 700 nm be absorbed, while below 700 nm be reflected into the light path. And the other side makes the halogen lamp irradiance below 700 nm be absorbed, while above 700 nm get through into the light path.

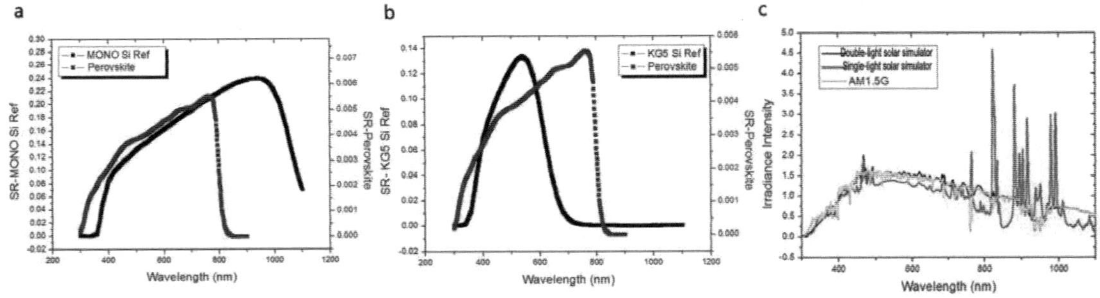

Figure 4. a) Spectral responsivity curves of mono-Si reference and a perovskite solar cell; b) Spectral responsivity curves of KG5-filter Si reference and a perovskite solar cell; c) Spectral distribution of double-light and single-light solar simulator compare to AM1.5G standard spectral irradiance.

Here for terrestrial solar cell, it is mean to match with AM1.5G. Table 2 displayed the MMF calculation results of the perovskite solar cell with different references and solar simulators, according to equation (1). Obviously, double-light solar simulator with mono-Si reference solar cell has the smallest spectral mismatch error, only -0.08%. It is the best partner of light source and reference solar cell. If use a mono-Si reference and a single-light solar simulators to measure the perovskite solar cell without modification, a deviation of 10.4% would be induced, so huge that cannot be ignored. Usually, if the mismatch error is smaller than 1%, no irradiance modification is needed, or else, irradiance modification should be conducted. That is, the original irradiance should be divided by the MMF value to obtain the corrected irradiance. Coordination between the light source and reference solar cells determines the MMF value. Table 2 also illustrated that the highly-matched double-light solar simulator is critical to reduce the mismatch error.

Table 2. (MMF-1) calculation results of the same perovskite solar cell measurement, by using NIM's different reference solar cells and solar simulators.

Reference solar cell	Double-light solar simulator	Single-light solar simulator
Mono-Si Ref	-0.08%	-10.4%
KG5-filtered Si Ref	1.0%	-1.5%

2.3. Other influencing factors

Above is to guarantee the effective traceability of the calibrated value, also to guarantee the first two conditions of STC that is 1000W/m^2 and AM1.5G. The third condition of STC is temperature of 25℃. Accurately speaking, it should be junction temperature of the measured PV device. Under the irradiation of solar simulator, solar cell's junction temperature will rapidly rise up inevitably. For the reference solar cell, we use a temperature controller to stable it to (25±1)℃, with an embedded Pt100 sensor to monitor the true state. But new type solar cells are always tiny and have no temperature sensor, its temperature coefficient is also usually unknown, so approximate temperature control would be conducted. If possible, temperature coefficient of new type solar cell should also be known.

Besides, *I-V* characteristics of new type non-silicon solar cells always influenced by scan speed, direction, etc., [19, 20]. For perovskite solar cells, its hysteresis is very common. In order to answer this problem, we developed software match to Keithly 2400 meter, which scan in both directions as well as measure the stabilized power output. Instability is another well-known performance of new type solar cells. They usually decay during measurement or even preservation, because of irradiance or/and humidity, and other factors. Only few of them are stable for a long time when exposed in air environment. In the present situation, scientists always report the highest results selectively in the literature. We hope the instability problems be solved early.

In the last, designated area of new type solar cells are also important for measurement accuracy. New type solar cells are usually tiny and transparent devices, stray light through non-effective area may add the total irradiance randomly. Masks should be taken into consideration, and its size is also critical for the efficiency measurement and calculation. Results between without mask and with masks whose area larger or smaller than cell itself, is vastly different. About the size-choosing of mask, there is no related formal standard yet. Empirically, take the sample's non-uniformity into consideration, we would employ a slightly smaller mask, and use the mask's calibrated area as the effective area for solar cell's efficiency calculation. The mask should be as thin as possible with regular shape. We have made precisely processed and calibrated mask in square and round shapes, with side length or diameter from 3mm to 10mm, which can cover most of the research cells' size. Besides measured *I-V* characteristic parameters, such as short circuit current, open circuit voltage, maximum power, efficiency, and so on, measurement conditions would be all demonstrated in the test report. That is reference solar cell's CV, scan direction, calibrated area of the mask, etc.

Figure 5a has shown the I-V curves of a perovskite solar cell measured by scanning in forward (red line) and reverse direction (blue line), critical photo-electric parameters can be read out from the curves, and efficiency can be calculated by the parameters, as listed below the curves, red corresponding to forward scan, blue corresponding to reverse scan. During these years, we have calibrated various kinds of new type non-silicon solar cells prepared by research institutes and universities. The highest efficiency record for various kinds of new type non-silicon solar cells were demonstrated in Figure 5b. In this scheme, the highest efficiency values measured in our laboratory during Jan. 2014 and Nov. 2017 are listed. As we can see, the highest efficiency record of organic solar cells is 13.9%. The highest efficiency record of pervoskite solar cells is 22.4%. Most of the results have been or will be published in famous journals.

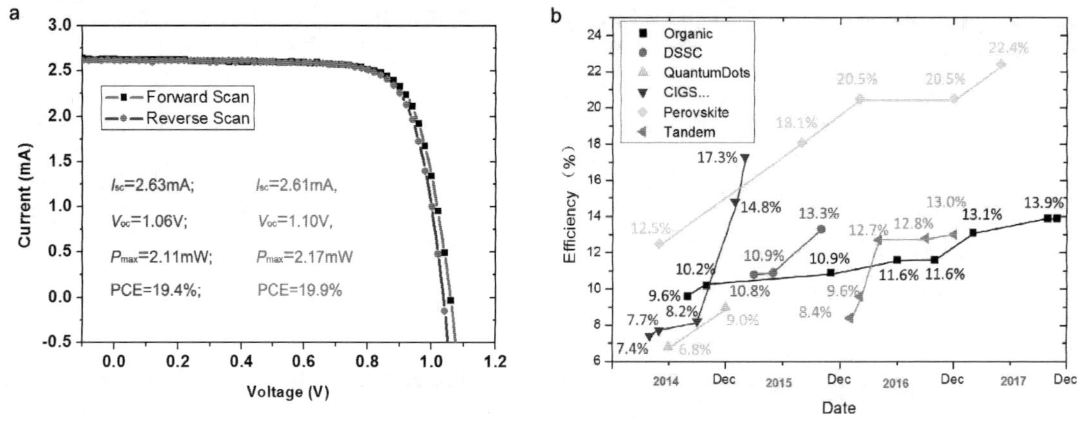

Figure 5. a) I-V curves of a perovskite solar cell by forward (Red) and reverse scan (Blue). b) Scheme of the measured highest efficiency record for various kinds of new type solar cells.

3. Conclusion

Based on the metrology traceability chain, we illustrated an accurate measurement method for new type non-silicon solar cells' photoelectric conversion efficiency. Taking the STC conditions and sample-related properties into consideration, main influencing factors were analyzed to reduce the measurement errors. Miscalculation and low comparability can be relieved or avoided by this method, it will provide strong support for the further research and application of new type solar cells.

Acknowledgments

We wish to acknowledge the assistance of National Key R&D Program of China (2016YFF0200801) and trust from research institutes and universities of China in new type solar cell measurement.

References

[1] Green M A et al. 2017 Solar cell efficiency tables (version 49) *Prog. Photovolt: Res. & Appl.*, **25** 144

[2] Li M et al. 2017 Solution-processed organic tandem solar cells with power conversion efficiencies >12% *Nature Photon.* 11

[3] Deng D et al. 2016 Fluorination-enabled optimal morphology leads to over 11% efficiency for inverted small-molecule organic solar cells *Nat. Commun.* **7** 13740

[4] Zhao W et al. 2017 Molecular Optimization Enables over 13% Efficiency in Organic Solar Cells *J. Am. Chem. Soc.* **139** 7148

[5] Jiang Q et al. 2016 Enhanced electron extraction using SnO_2 for high-efficiency planar-structure $HC(NH_2)_2PbI_3$-based perovskite solar cells *Nature Energy* **2** 16177

[6] Pan Z et al. 2014 High-efficiency "green" quantum dot solar cells *J. Am. Chem. Soc.* **136** 9203

[7] Snaith H J 2012 The perils of solar cell efficiency measurements *Nature Photon.* **6** 337

[8] Zimmermann E et al. 2014 Erroneous efficiency reports harm organic solar cell research, *Nature Photon.* **8** 669

[9] Grätzel M 2014 The light and shade of perovskite solar cells *Nature Mater.* **13** 838

[10] Meng H et al. 2016 Research advancement of measurement of new type solar cells' photo-electric conversion efficiency *Imaging Science and Photochemistry* **34** 389

[11] Ye L et al. 2015 Toward reliabl e and accurate evaluation of polymer solar cells based on low band gap polymers, *J. Mater. Chem. C* **3** 564

[12] 2006 IEC 60904-1 Photovoltaic Devices Part 1 Measurement of Photovoltaic Current-Voltage Characteristics

[13] Meng H et al. 2014 Calibration of solar cells' photoelectric properties and related uncertainty analysis *Proc. of SPIE* **9233** 92331T

[14] 2008 IEC 60904-7 (Ed. 3.0) Photovoltaic devices - Part 7: Computation of the spectral mismatch correction for measurements of photovoltaic devices

[15] 2009 IEC 60904-4 (Ed. 1.0), Photovoltaic devices – Part 4: Reference solar devices – Procedures for establishing calibration traceability

[16] He Y et al. 2015 Primary calibration of solar cells based on DSR method at the National Institute of Metrology of China *Proc. of SPIE* **9623** 96230S-1

[17] 2007 IEC 60904-9 Photovoltaic Devices: Part 9. Solar Simulator Performance Requirements

[18] Meng H et al. 2011 Research on integrated system for solar simulator performance calibration according to IEC 60904-9 *Proc. of SPIE* **8201** 82012L-1

[19] Yang X et al. 2012 Reliable evaluation of dye-sensitized solar cells, *Energy Environ. Sci.* **6** 54

[20] Snaith H J et al. 2014 Anomalous Hysteresis in Perovskite Solar Cells, *J. Phys. Chem. Lett.* **5** 1511

IOP Conf. Series: Journal of Physics: Conf. Series **972** (2018) 012018 doi:10.1088/1742-6596/972/1/012018

Traceable calibration of photovoltaic reference cells using natural sunlight

H Müllejans, W Zaaiman, D Pavanello and E D Dunlop

European Commission, Joint Research Centre, Directorate C – Energy, Transport & Climate, Energy Efficiency and Renewables Unit, European Solar Test Installation (ESTI), Via Fermi 2749, I-21027 Ispra (VA), Italy

E-mail: harald.muellejans@ec.europa.eu

Abstract. At the European Solar Test Installation (ESTI) photovoltaic (PV) reference cells are calibrated traceably to SI units via the World Radiometric Reference (WRR) using natural sunlight. The Direct Sunlight Method (DSM) is described in detail and the latest measurement results and an updated uncertainty budget are reported. These PV reference cells then provide a practical means for measuring the irradiance of natural or simulated sunlight during the calibration of other PV devices.

1. Introduction

The price of PV modules is directly determined by their electrical performance at standard test conditions (STC). Therefore the independent determination of the electrical performance of PV modules with high accuracy (i.e. small measurement uncertainty) is important to guarantee fair and transparent markets. The (inevitably) remaining uncertainty contributes to the financial risk of PV module production and installation.

The electrical performance of PV modules is normally measured under pulsed simulated sunlight. The most difficult parameter to control and measure is the irradiance of the light (which should be 1000 W/m^2 at STC), whereas other quantities such as current and voltage are more readily available. PV reference cells are commonly used for the measurement of the irradiance of the simulated sunlight, as these have similar characteristics (in particular in response time) to the PV modules. The challenge is to calibrate these devices as irradiance sensors traceably with the lowest possible uncertainty. At ESTI this is achieved by calibration under natural sunlight against cavity radiometers, which are secondary standards for solar irradiance [1], directly traceable to the World Standard Group (WSG), which is the international primary standard for solar irradiance. Therefore traceability is to the World Radiometric Reference (WRR) and thereby SI units. This paper presents the DSM as implemented at ESTI and results obtained therewith.

2. Methods

The calibration of PV reference cells against natural sunlight can either use only the direct normal solar irradiance or alternatively the global in plane solar irradiance. In the former (DSM) only the direct sunlight (including the circumsolar) is used by excluding the other components (diffuse irradiance from the sky as well as reflected from the ground). This is achieved by suitable collimation tubes. The Global Sunlight Method (GSM) does not use such collimation tubes and therefore all components of the irradiance contribute to the measurement. All methods are described in general terms in IEC 60904-4 [1]. The calibration has to be reported at STC which prescribes an irradiance of 1000 W/m^2 with a global spectral irradiance distribution according to IEC 60904-3 [2] and a device junction temperature of 25°C. Measurements are taken at conditions close to STC and appropriately corrected as described below.

2.1. Direct Sunlight Method (DSM)

The DSM has been used traditionally by NREL [3], and subsequently was first implemented at ESTI in 2005 [4]. In the meanwhile the set-up and method has been further refined and is described in detail here. In brief the direct solar irradiance is measured, the total with cavity radiometers and its spectral distribution with a suitable spectroradiometer. Furthermore the short circuit current of the reference cells is measured in parallel with the above mentioned parameters, while the cells are temperature controlled. The field-of-view of both the spectroradiometer and the reference cells is limited to the same angle as the cavity radiometers by suitable collimation tubes.

2.1.1. Instrumentation and mounting. The direct irradiance was measured with three cavity radiometers (PMO6 nr. 81109 and PMO6 nr. 911204 and TMI IV nr. 68835) directly traceable to the WRR via the International Pyrheliometer Intercomparison (IPC) [5]. The cavity radiometers have a field of view full angle of 5°. The spectral irradiance was measured with a spectroradiometer (OL 750) equipped with an integrating sphere and an additional collimation tube. The instrument was calibrated in-house against a standard lamp traceable to the SI irradiance scale via NPL. The PV reference cells were mounted on a water cooled mounting plate which allowed to maintain the reference cell temperature at (25±2)°C during measurements. Collimating tubes were mounted in front of the cells. All equipment was mounted on solar trackers automatically following the sun. All collimation tubes were designed to limit the field of view for the spectroradiometer and the reference cells to the same as it is for the cavity radiometers. Therefore all instruments are exposed to the direct solar irradiance only. The set-up is shown in figure 1 and figure 2.

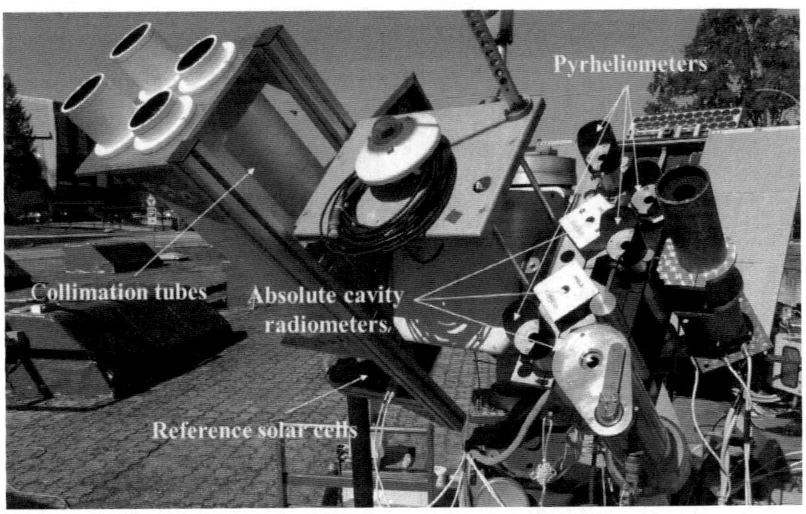

Figure 1: The main set-up with the three cavity radiometers and reference solar cells below the collimation tubes. Further auxiliary instrumentation is also shown but not used for the DSM except for one of the pyrheliometers for monitoring the stability of solar irradiance.

Figure 2: The OL750 spectroradiometer with collimation tube attached to the integrating sphere.

2.1.2. Measurements. The two PMO6 instruments were operated on a 40 second cycle time, giving a data point every 80 seconds (40 seconds open and 40 seconds closed). The TMI instrument was read every 20 seconds. The short circuit current of the reference cells was recorded simultaneously with a multi-channel transimpedance amplifier and data logger together with the reading from the internal temperature sensor to the reference cells every 20 sec. The OL750 spectroradiometer was operated in automatic mode (using a suitable script) starting a new acquisition every 10 minutes approximately.

2.1.3. Irradiance correction. The following stability criteria were chosen: irradiance from the two PMO6 cavities within 1 W/m^2, irradiance stable to within ± 0.5% over 80 seconds (PMO6 cycle time), irradiance stable to within ± 1.0% over 300 seconds (OL750 cycle time). The irradiance stability is taken from the reading of a pyrheliometer (ser. Nr. 930018), which has a faster response time than the cavities (and is read every 20 sec). For data points fulfilling these criteria, the direct irradiance G_{dir} is taken as the average reading of all three cavities and finally accepted if larger than 750 W/m^2. A systematic shift between WRR and SI scale has been established [6]. Therefore the readings of the cavity radiometers were corrected to the SI irradiance scale by dividing the WRR reading by 1.0034. During stable conditions, an irradiance reading is therefore available every 80 seconds. The four short circuit current readings (20 sec) for each irradiance reading (80 sec) are averaged. The short circuit current scaled to 1000 W/m^2 (I_{SC}) is then calculated from the measured short circuit current I_{meas} by

$$I_{SC} = I_{meas} \frac{1000 \frac{\mathrm{W}}{\mathrm{m}^2}}{G_{dir}} \tag{1}$$

2.1.4. Temperature correction. Considering the narrow temperature range no correction for cell temperature was made. The deviation of up to 2°C from the target temperature of 25°C is treated as a measurement uncertainty component and explicitly included in the respective calculation.

2.1.5. Spectral correction. The direct spectral irradiance data from the OL750 were visually inspected and all spectra which showed signs of instability or artefacts were eliminated. The acquisition of a single spectrum takes 5 to 6 minutes, and is sequential in wavelength, starting at 300 nm up to 2500 nm. Instable conditions (e.g. moving clouds) lead to decreased or increased signal in some parts of the spectrum, thereby distorting its shape. They are easily identified by visual inspection comparing with the daily average and the reference spectrum [2]. A spectral correction was applied for each data point according to

$$CV = I_{SC} \frac{\int SR_{DUT}(\lambda) E_{ref}(\lambda) \mathrm{d}\lambda}{\int SR_{DUT}(\lambda) E_{meas}(\lambda) \mathrm{d}\lambda} \frac{\int E_{meas}(\lambda) \mathrm{d}\lambda}{\int E_{ref}(\lambda) \mathrm{d}\lambda} \tag{2}$$

with CV the calibration value at STC, I_{SC} from equation (1), E_{meas} the measured spectral irradiance, E_{ref} the reference global spectral irradiance [2], SR_{DUT} the spectral responsivity of the reference cell and $\boldsymbol{\lambda}$ the wavelength.

2.1.6. Integration range. The data were interpolated and then integrated from 300 nm to 2500 nm. This introduces a systematic error, as the integration in equation (2) should be over all wavelengths, i.e. from zero to infinity. The lower integration limit of 300 nm does not introduce any effect as the spectral irradiance below this wavelength can be neglected [2]. However, above 2500 nm there are 10.3 W/m^2 in both, the global and the direct reference spectra [2]. This affects only the two integrals in the second fraction in equation (2), as the SR of reference cells vanishes above 1200 nm. The irradiance missing from the two integrals is the same in absolute terms for global and direct spectra, but has a different relative contribution as the total irradiances of the two spectra are 1000 W/m^2 and 900 W/m^2 respectively. Therefore a correction factor of 1.00116 was calculated (approximating the measured spectra with the direct reference spectrum) and applied.

2.1.7. Averaging. The data were averaged [1] as summarized below. For each acquired spectrum typically seven I_{SC} values were available, sometimes less due to the described filtering. All available values per spectrum were averaged, but if there were less than four data, the point was eliminated. Then all results from each day were averaged with a minimum of five data points per day. Finally the results of all days were averaged with a minimum of at least three days for a valid *CV* as required by IEC 60904-4 [1].

2.2. Global Sunlight Method (GSM)
The GSM has already been described extensively in the past [7].

3. Results
Here we report the results of the two most recent calibration campaigns in 2014 and 2015. In 2014 five ESTI PV reference cells (PX201C, PX301C, PX304C, 930417-2 and ASP010) were calibrated on 31 of July, 4 and 21 of August 2014. In 2015 two of the former (PX201C and PX301C) were re-calibrated on 20, 21 and 22 of April 2015. The other cells were also measured in 2015 but it was not possible to obtain data on three different days, and therefore no full calibration was achieved. All reference cells are crystalline silicon.

3.1. Spectral irradiance
Measured spectra of direct solar irradiance for 4.8.14 and 20.4.2015 are plotted in figure 3 and figure 4. A certain spectral variation during the day is observed due to the change of the path length of the solar direct beam within the atmosphere. All measured spectra deviate from but are similar to the reference spectrum. This was analysed by classifying all spectra according to IEC 60904-9 [8]. This showed that all spectra were meeting the requirements for class A and 95% of all spectra had deviations from the reference spectrum less than half the allowed limit specified for class A. The analysis in IEC 60904-9 [8] is limited to the wavelength range 400 nm to 1100 nm. However, this corresponds to where the spectral responsivity of the reference cells is significant. Therefore, it gives a good indication that for the reference cells investigated here the spectrum of natural sunlight is similar to the reference spectrum. The deviations from the latter are corrected for with equation (2).

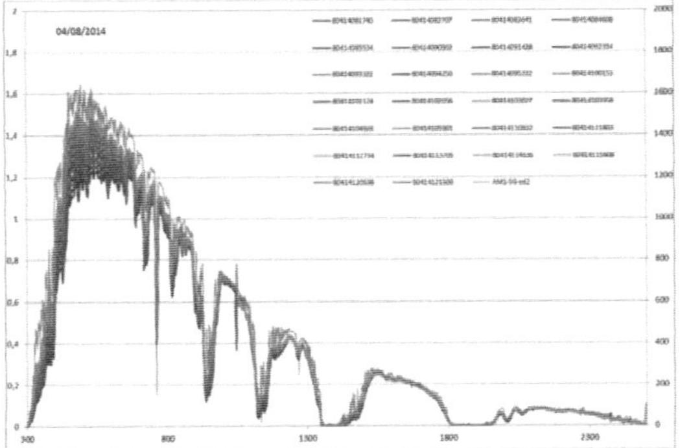

Figure 3: Measured direct solar spectral irradiance on 4.8.14 with reference spectrum (in red).

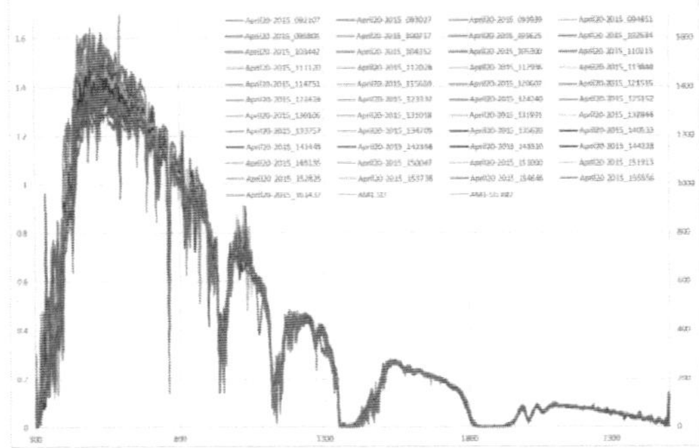

Figure 4: Measured direct solar spectral irradiance on 20.4.15 with reference spectrum (in red).

3.2. Short circuit current and calibration values

As an example the data analysis for cell PX301C in 2015 is shown in figure 5. Table 1 lists all results from both campaigns, for single days as well as the final averaged *CV*. The spread of the data for cell PX201C in 2015 is somewhat larger due to limited (but still sufficient) number of data points.

3.3. Measurement uncertainty UC

The UC budget for DSM and GSM has been updated after the shift of the WRR to SI irradiance scale had been implemented. Previously the uncertainty of scale difference had been estimated to be 0.3% with rectangular distribution (i.e. 0.17% $k=1$). The recently established scale shift has a lower uncertainty of 0.09% ($k=1$). Thereby the overall combined expanded uncertainty of the DSM method is 0.48% ($k=2$) (table 2). The respective uncertainty for the *CV* of each of the reference cells is included in table 1.

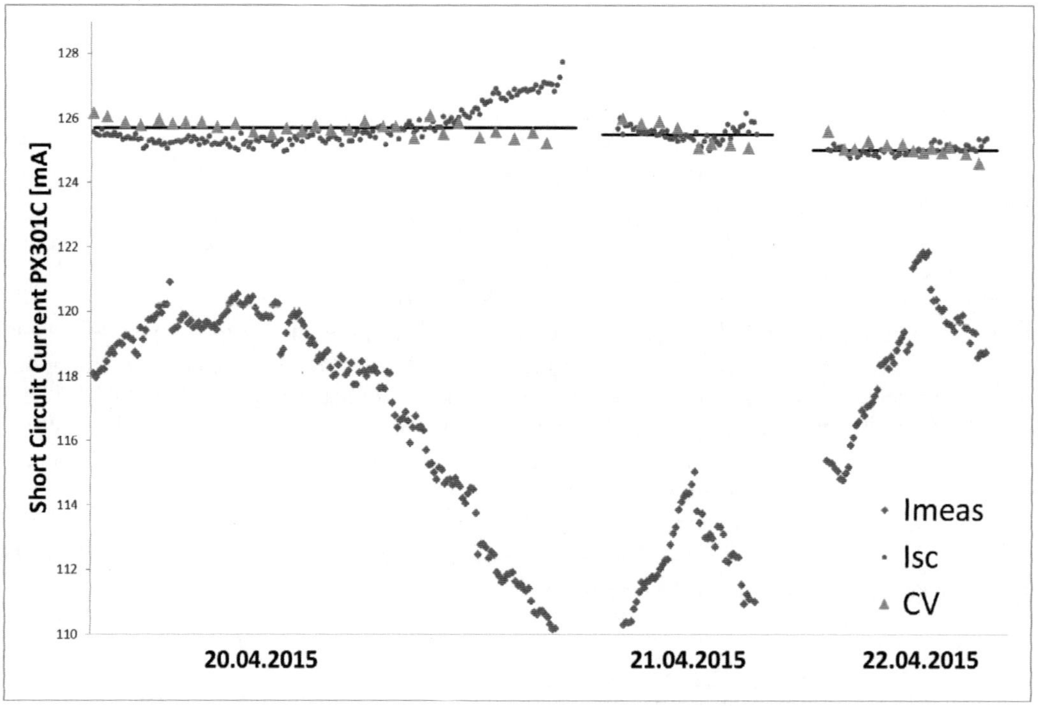

Figure 5: Data analysis for reference cell PX301C in 2015. The blue squares are the as measured short circuit current I_{meas}, the red dots the I_{SC} (scaled to 1000 W/m^2) and the green triangles the results after spectral correction. The black horizontal lines are the daily averages calculated from the latter.

Table 1: The results for all cells in 2014 and 2015 for the single days and the average. All values are in mA.

	ASP010	PX201C	PX301C	PX304C	930417-2		PX201C	PX301C
31.07.14	79.81	123.43	125.45	124.31	123.80	20.04.15	123.90	125.85
04.08.14	79.87	123.65	125.67	124.57	123.94	21.04.15	124.14	125.63
21.08.14	79.75	123.35	125.36	124.26	123.65	22.04.15	122.95	125.17
Average	**79.81**	**123.48**	**125.49**	**124.38**	**123.80**	**Average**	**123.67**	**125.55**
UC ($k=2$)	0.34	0.59	0.60	0.60	0.59	UC ($k=2$)	0.59	0.60

Table 2: Uncertainty budget for DSM and GSM against SI units before (WRR) and after (SI) shift of the WRR irradiance scale. All contributions are listed as standard UCs in %.

Uncertainty component	DSM (WRR)	DSM (SI)	GSM (WRR)	GSM (SI)
Data acquisition	0.06	0.06	0.06	0.06
Temperature effects	0.06	0.06	0.06	0.06
Irradiance	0.11	0.11	0.24	0.24
Spectral correction	0.18	0.18	0.18	0.18
WRR offset to SI	0.17	0.09	0.17	0.09
Combined standard UC ($k=1$)	0.29	0.24	0.35	0.32
Combined expanded UC ($k=2$)	**0.58**	**0.48**	**0.70**	**0.64**

4. Summary and conclusions

The DSM as implemented at ESTI provides calibration of PV reference cells under natural direct solar irradiance against secondary standards for solar irradiance and is traceable to SI units via the WRR. The expanded UC is below 0.5% ($k=2$). Here the results obtained in 2014 and 2015 for five different cells are reported. The cells are then used to measure the irradiance of natural and simulated sunlight during the calibration of electrical performance of other PV devices, providing the traceability for these measurements. Furthermore the results contribute to the determination of key comparison reference value (KCRV) for each reference cell, reducing further their uncertainty and establishing the World PV Scale (WPVS) [9].

Acknowledgements
We acknowledge contributions from our colleagues at ESTI.

References

[1] IEC 60904-4 2009 "Photovoltaic devices - Part 4: Reference solar devices - Procedures for establishing calibration traceability".
[2] IEC 60904-3 2016 "Photovoltaic devices - Part 3: Measurement principles for terrestrial photovoltaic (PV) solar devices with reference spectral irradiance data".
[3] Osterwald C R, Emery K A, Myers D R and Hart R E 1990 "Primary reference cell calibrations at SERI: History and methods" *Proc. 21st IEEE PVSC* 1062-7.
[4] Zaaiman W, Müllejans H, Dunlop E D and Ossenbrink H A 2006 "SI traceability for irradiance transfer standards to silicon solar cells using direct and global irradiance" *Proc. 21st EU PVSEC* 64-8.
[5] Finsterle W 2016 "International Pyrheliometer Comparison IPC-XII" WMO IOM Report No. 124.
[6] Fehlmann A, Kopp G, Schmutz W, Winkler R, Finsterle W and Fox N 2012 "Fourth world radiometric reference to SI radiometric scale comparison and implications for on-orbit measurements of the total solar irradiance" *Metrologia* **49** S34-8.
[7] Müllejans H, Zaaiman W, Dunlop E D and Ossenbrink H A 2005 "Calibration of photovoltaic reference cells by the global sunlight method" *Metrologia* **42** 360-7.
Müllejans H, Ioannides A, Kenny R, Zaaiman W, Ossenbrink H A and Dunlop E D 2005 "Spectral mismatch in calibration of photovoltaic reference devices by global sunlight method" *Measurement Science and Technology* **16** 1250-4.
[8] IEC 60904-9 2007 "Photovoltaic devices - Part 9: Solar simulator performance requirements".
[9] Müllejans H, Zaaiman W and Dunlop E D 2015 "Reduction of uncertainties for photovoltaic reference cells" *Metrologia* **52** 646-53.

IOP Conf. Series: Journal of Physics: Conf. Series **972** (2018) 012019 doi:10.1088/1742-6596/972/1/012019

Calorimetric Measurement for Internal Conversion Efficiency of Photovoltaic Cells/Modules Based on Electrical Substitution Method

Terubumi Saito[1, 2], Muneaki Tatsuta[1], Yamato Abe[1] and Minato Takesawa[2]

[1]Tohoku Institute of Technology, Sendai, Miyagi 982-8577, Japan
[2]Graduate School, Tohoku Institute of Technology, Sendai, Miyagi 982-8577, Japan

terubumi.saito@tohtech.ac.jp

Abstract. We have succeeded in the direct measurement for solar cell/module internal conversion efficiency based on a calorimetric method or electrical substitution method by which the absorbed radiant power is determined by replacing the heat absorbed in the cell/module with the electrical power. The technique is advantageous in that the reflectance and transmittance measurements, which are required in the conventional methods, are not necessary. Also, the internal quantum efficiency can be derived from conversion efficiencies by using the average photon energy. Agreements of the measured data with the values estimated from the nominal values support the validity of this technique.

1. Introduction

Among the characteristics on solar cells/modules, power conversion efficiency is essentially important. The external power conversion efficiency is defined as the ratio of the electrical power generated to the radiant power incident to the solar cell/module under the IEC Standard Test Conditions [1]. While the evaluation for the output, the electrical power, can be conducted very easily, the evaluation for the input, the radiant power, typically needs defined measurements and corrections [1] based on a standard cell using specialized instruments, and therefore it can be done only at limited laboratories [2].

From the viewpoint of material physics, the internal conversion efficiency, which is defined by the electrical power generated to the radiant power absorbed, is more important than the external one. In addition, there exists needs for efficiency measurement technique usable under spectrally and/or angularly diverted conditions (for instance, under oblique incidence condition) from the Standard Test Conditions.

To meet such needs, we aim to directly determine the internal conversion efficiency by utilizing the electrical substitution technique for solar cells/modules under test by applying forward bias current to the test cells/modules, and to evaluate if the technique is feasible. The technique to calorimetrically determine the internal quantum efficiency was developed by T. Inoue et al. for photodiodes receiving a laser beam [3]. This is a first attempt to our knowledge to apply the technique to photovoltaic modules receiving non-monochromatic radiation to determine directly the internal conversion efficiency and derive the external one together with internal/external quantum efficiencies.

2. Principle of Operation

In our technique, the absorbed radiant power is determined by a calorimetric method in which the heat absorbed in the solar module is substituted by the electrical power applied to the solar cell as

IOP Conf. Series: Journal of Physics: Conf. Series **972** (2018) 012019 doi:10.1088/1742-6596/972/1/012019

illustrated in Figure 1. When the module is illuminated as shown on the left, a part of the incident radiant power, P_r, is reflected by (and transmitted through) the semiconductor substrate and the rest power is absorbed by the substrate. In the case when the module is open or shorted, all the absorbed power, P_a, is converted to the heat, P_h. In the case when the module is connected to a load, a part of the absorbed power is converted to the electricity, P_m, and the rest power generates the heat, P_h. When the incident radiation is shielded as shown on the right, the module temperature goes down. In this dark condition, the electrical power is applied from a power supply and is adjusted so that the module temperature becomes equal to the one in an illuminated condition. We can assume that the substituted electrical power is equal to the thermal power, or the absorbed radiant power in the case of an open- or short-circuit condition. When the load is connected, the absorbed radiant power is equal to P_h+P_m, the sum of the substituted electrical power and the power supplied to the load.

Since the maximum electrical power generated can be easily measured, the internal conversion efficiency is directly determined. External conversion efficiency and internal/external quantum efficiencies can also be obtained as we will see in the next section.

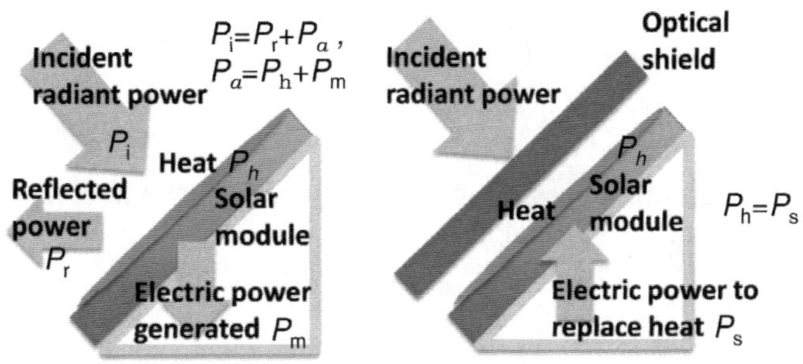

Figure 1. Schematics to show the electrical substitution technique for solar modules.

3. Analysis

3.1. Average photon energy

Figure 2 shows incident and absorbed spectra for (a) energy and (b) photons together with spectral absorptance curve. Average photon energy is necessary to convert between conversion efficiencies and quantum efficiencies. For radiation with the standard AM1.5 solar spectrum, the average incident photon energy is defined as the ratio of the spectral integration over wavelength, λ, of incident spectral irradiance, $E_i(\lambda)$ [W/(nm m^2)], to that of incident spectral photon irradiance, $X_i(\lambda)$ [1/(s nm m^2)], and derived as

$$a_i = \frac{\int_{\lambda_1}^{\lambda_2} E_i(\lambda)d\lambda}{\int_{\lambda_1}^{\lambda_2} X_i(\lambda)d\lambda} = \frac{hc\int_{\lambda_1}^{\lambda_2} E_i(\lambda)d\lambda}{\int_{\lambda_1}^{\lambda_2} \lambda E_i(\lambda)d\lambda} \approx 1.45 \text{ eV},$$

where λ_1=250 nm, λ_2=4000 nm, h the Planck constant, and c the speed of light in vacuum.

For the absorbed radiation, the average photon energy becomes a function of the absorption coefficient, $\alpha(\lambda)$, and of the substrate thickness, t. If the reflectance is neglected, it is defined as

$$a_a = \frac{\int_{\lambda_1}^{\lambda_2} E_i(\lambda)[1-\exp\{-\alpha(\lambda)t\}]d\lambda}{\int_{\lambda_1}^{\lambda_2} X_i(\lambda)[1-\exp\{-\alpha(\lambda)t\}]d\lambda} = \frac{hc\int_{\lambda_1}^{\lambda_2} E_i(\lambda)[1-\exp\{-\alpha(\lambda)t\}]d\lambda}{\int_{\lambda_1}^{\lambda_2} \lambda E_i(\lambda)[1-\exp\{-\alpha(\lambda)t\}]d\lambda}.$$

For the standard AM1.5 solar spectrum and 0.4 mm-thick crystalline Si, it is evaluated by using the published optical constants [4] to be

$$a_a \approx 1.89 \text{ eV}.$$

Spectrally integrated energy absorptance is defined and calculated as,

$$A_e = \frac{\int_{\lambda_1}^{\lambda_2} E_i(\lambda)[1-\exp\{-\alpha(\lambda)t\}]d\lambda}{\int_{\lambda_1}^{\lambda_2} E_i(\lambda)d\lambda} \approx 0.773,$$

while spectrally integrated photon absorptance is defined and calculated as,

$$A_{\mathrm{p}} = \frac{\int_{\lambda_1}^{\lambda_2} X_{\mathrm{i}}(\lambda)[1-\exp\{-\alpha(\lambda)t\}]d\lambda}{\int_{\lambda_1}^{\lambda_2} X_{\mathrm{i}}(\lambda)d\lambda} = \frac{\int_{\lambda_1}^{\lambda_2} \lambda E_{\mathrm{i}}(\lambda)[1-\exp\{-\alpha(\lambda)t\}]d\lambda}{\int_{\lambda_1}^{\lambda_2} \lambda E_{\mathrm{i}}(\lambda)d\lambda} \approx 0.592.$$

Among the four values, the following relationship holds.

$$a_{\mathrm{i}}A_{\mathrm{e}} = a_{\mathrm{a}}A_{\mathrm{p}}$$

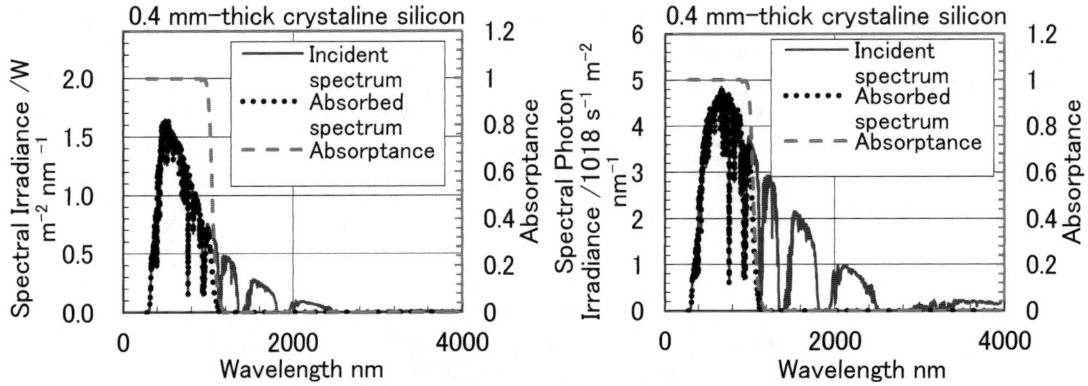

Figure 2. (a) Incident and absorbed spectral irradiances. (b) Incident and absorbed spectral photon irradiances. Spectral absorptance curve is also shown.

3.2. Experimental efficiencies derivation

Definitions [5,6] for power conversion efficiencies and quantum efficiencies are tabulated in Table 1.

Table 1. Definitions for power conversion efficiencies and quantum efficiencies

	Conversion efficiency	Quantum efficiency
External	$C_{\mathrm{ext}} = \frac{P_{\mathrm{m}}}{P_{\mathrm{i}}} = A_{\mathrm{e}}C_{\mathrm{int}}$	$\eta_{\mathrm{ext}} = \frac{nI_{\mathrm{sc}}}{e\Phi_{\mathrm{i}}} = A_{\mathrm{p}}\eta_{\mathrm{int}}$
Internal	$C_{\mathrm{int}} = P_{\mathrm{m}}/P_{\mathrm{a}}$	$\eta_{\mathrm{int}} = \frac{nI_{\mathrm{sc}}}{e\Phi_{\mathrm{a}}} = \frac{na_{\mathrm{a}}I_{\mathrm{sc}}}{P_{\mathrm{a}}}$

Here, notations are as follows. P_a [W]: absorbed radiant power, Φ_a [1/s]: absorbed photon flux, n: number of solar cells in the module, P_{m} [W]: maximum electrical power generated, I_{sc} [A]: short circuit current, e [C]: elementary electronic charge.

In our technique, the absorbed radiant power can be directly determined as explained. Therefore, the internal conversion efficiency is, in the first place, determined based on the definition. Secondly, the external power conversion efficiency was derived from the internal power conversion efficiency using the calculated absorptance. Lastly, the internal/external quantum efficiencies were converted from the internal/external power conversion efficiencies by using the short circuit current and the average photon energy as shown in Table 1.

3.3. Efficiencies estimated from nominal values

To compare the measured efficiencies, conversion efficiencies and quantum efficiencies for a module and a cell estimated from the nominal values were calculated as shown in Table 2. Here, E=1000 [W/m^2]: the irradiance of STC, S [m^2]: module area, S_{c} [m^2]: cell area, n: number of cells per module,

Table 2. Equations to derive various efficiencies from nominal values

	Conversion efficiency		Quantum efficiency	
	module	cell	module	cell
External	$C_{\text{ext,m}} = \frac{P_{\text{m}}}{ES}$	$C_{\text{ext,c}} = \frac{P_{\text{m}}}{nES_{\text{c}}}$	$\eta_{\text{ext,m}} = \frac{na_{\text{i}}I_{\text{sc}}}{ES}$	$\eta_{\text{ext,c}} = \frac{a_{\text{i}}I_{sc}}{ES_{\text{c}}}$
Internal	$C_{\text{int,m}} = \frac{C_{\text{ext,m}}}{A_{\text{e}}}$	$C_{\text{int,c}} = \frac{C_{\text{ext,c}}}{A_{\text{e}}}$	$\eta_{\text{int,m}} = \frac{\eta_{\text{ext}}}{A_{\text{p}}}$	$\eta_{\text{int,c}} = \frac{\eta_{\text{ext,c}}}{A_{\text{p}}}$

4. Experiments

Measurements were carried out on two types of crystalline silicon solar modules, specifications of which are tabulated in Table 3. Two sets of identical solar modules were set in the same outdoor conditions with the appropriate load (5.29 Ω for HIP-63S1, 46.6 Ω for GT-833S-TF) to attract maximum power. In addition, the electrical substitution experiments were conducted also for the modules (GT-833S-TF) in an open- and short-circuit conditions. Module temperatures were monitored by thermocouples attached to the center of the rear side of the modules and also by a thermographic camera set behind the modules.

Experiment procedures are as follows.

1. Align the modules to the sun and wait until the thermal equilibrium is established.
2. Shade one (test module) of the two modules completely with a sun shield.
3. Apply electrical power (in constant current mode) to the test module and adjust it so that its temperature becomes equal to the temperature of the other module (reference module).
4. Take the substituted electrical power plus the electrical power generated when the sun is not shielded as the absorbed power. Analyze the data.

Absorptance of the substrate is evaluated by the calculation using the literature absorption coefficient of crystalline silicon as described in section 3.1.

Table 3. Nominal values and specifications of photovoltaic modules tested.

Manufacturer	SANYO	KIS
Model	HIP-63S1	GT-833S-TF
Type	HIT	c-Si
Nominal maximum power /W	63	7
Nominal short circuit current /A	3.75	0.45
Nominal open circuit voltage /V	22.6	21.1
Number of cells	32	34
Module area /m^2	0.465	0.0466
Cell area /m^2	0.011	0.00123

5. Results

Figure 3 shows one of the measurement results (done in the campus on 2014/10/25) for the HIT type modules (Sanyo HIP-63S1) as a function of time. The test module was sun-shielded from 11:55 until the end of the graph. Even in the sunny condition, the temperatures of both modules fluctuated mainly due to the wind. The small difference in temperature between the modules at the same condition is also likely to be attributed to the difference in the wind condition.

Electrical power (97 W) started to be applied to the test module at 12:11. The applied power was increased to around 200 W at 12:20. With this power, both temperatures of the test module and the reference one agree. In this duration, the average substituted power was 203 W while the averaged maximum electrical power generated from the reference module was 61.3 W. It means that the sum, 264.3 W, is equal to the absorbed power. Therefore, the internal conversion efficiency is simply given by the ratio of 61.3 W to 264.3 W resulting in 0.232 as shown in Table 4 together with other efficiencies and comparison with the estimations from nominal values. For all the measured efficiencies, spectral mismatch corrections [7], which are required for rigorous evaluation to take account of the difference between the actual spectrum and the standard one were not applied.

In Table 4, separate measurement results on a different day (2014/10/29) are also shown to check the reproducibility. All the second measurement results are smaller than the first one by 2.2%, 2.2%, 1.9% and 1.8% for ICE, ECE, IQE and EQE, respectively. Reproducibilities (k=2) obtained are 3.1%, 3.2%, 2.7% and 2.6% for each.

Concerning the comparison with the efficiencies from nominal values, all the measured results are closer to the corresponding cell efficiency than the module efficiency; the measured efficiencies are smaller by 4-6% than the cell efficiencies while they are larger by 30-33% than the module efficiencies. Although the uncertainty is in the process to be analyzed, a tentative uncertainty is estimated to be 5-6% (k=2). The result that the measured efficiencies are closer to the cell efficiencies can be understood reasonable considering that radiation heating is rather confined to the area of the cell substrate because of the whiteness of the back-sheet, which is similar to the electric joule heating.

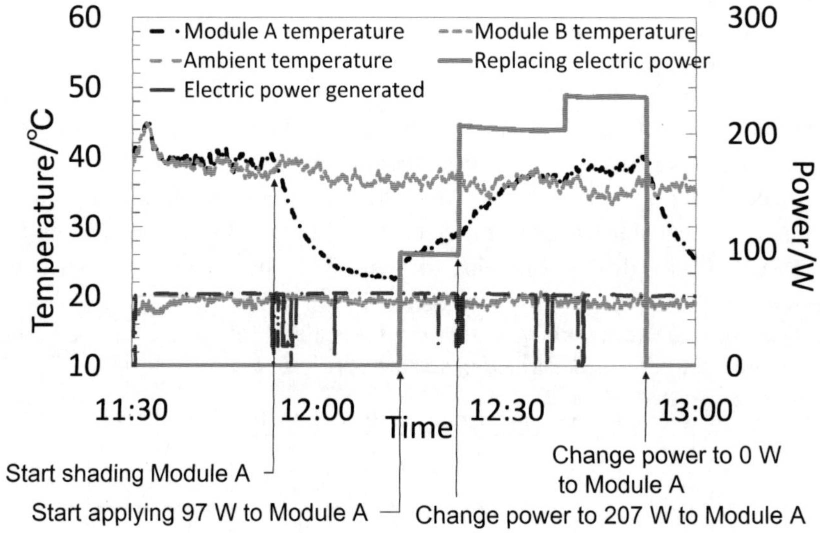

Figure 3. Time chart for measured data and applied electrical power.

Table 4. Comparison on Sanyo HIP-63S1 between experimentally determined efficiencies and estimated ones from nominal values. ICE / ECE: Internal / External Conversion Efficiency, IQE / EQE: Internal / External Quantum Efficiency.

Sanyo HIP-63S1		ICE	ECE	IQE	EQE
Estimated from nominal values	Module	0.175	0.135	0.632	0.374
	Cell	0.242	0.187	0.874	0.517
Measured (Loaded)	2014/10/25	**0.232**	**0.179**	**0.837**	**0.495**
	2014/10/29	**0.227**	**0.175**	**0.821**	**0.486**

Table 5 shows additional results on different smaller crystalline silicon modules (KIS GT833-TF) to compare efficiencies under loaded, open, and shorted conditions together with estimated ones efficiencies from nominal values. Regardless of the load conditions, the conversion efficiencies agree within ±4.8% (k=2) and the quantum efficiencies agree within ±2.5% (k=2). However, contrary to the results in Table 4, most of the measured efficiencies are closer to the module efficiencies. It is highly likely for this type of small modules that the nominal values such as maximum power etc. are not reliable enough but has high uncertainties.

Table 5. Comparison on KIS GT833-TF between experimentally determined efficiencies under loaded, open, shorted and estimated ones from nominal values.

KIS GT833-TF		ICE	ECE	IQE	EQE
Estimated from nominal values	Module	0.194	0.150	0.804	0.476
	Cell	0.216	0.167	0.894	0.530
Measured	Loaded	**0.188**	**0.145**	**0.731**	**0.433**
	Opened	**0.195**	**0.151**	**0.722**	**0.428**
	Shorted	**0.207**	**0.160**	**0.758**	**0.449**

To automatize the electrical substitution, we have also conducted experiments with negative feedback control system, which automatically adjusts the applied electrical power to maintain a constant temperature. The automatic substitution system also has been proved to work successfully and is to be published somewhere.

6. Conclusions

It has been proven that internal power conversion efficiencies can be successfully determined by a calorimetric method by which the absorbed radiant power can be determined by replacing the heat absorbed in the solar module with the electrical power. The technique is advantageous in that the reflectance and transmittance measurements, which are required in the conventional methods, are not necessary. The internal power conversion efficiencies are converted to external one by using the absorptance of the solar module. Also, the internal and external quantum efficiencies can be derived from conversion efficiencies by using the average photon energy. Agreements of the measured data with the values estimated from the nominal values support the validity of this technique.

ECEs and EQEs obtained by this technique are compared with those derived from nominal values for module and cell efficiencies. All the reliable measured efficiencies lie between the estimated cell efficiency and the module one and are closer to the former. This is reasonable considering the white back-sheet reflects most of the incident radiation and therefore generates little heat outside of the substrate, which is similar to the joule heating by electricity.

The disagreement are considered to be caused by the following factors: temperature fluctuation due to wind, inequivalence between radiation heating and electrical heating, uncertainty in nominal values of the solar modules, deviation from the standard test conditions and so on. Further study is being performed to investigate the details and to improve the performance, for instance, by introducing the automatic substitution method.

References

[1] IEC 60891: 2009 Ed 2 Procedures for temperature and irradiance corrections to measured I-V characteristics of crystalline silicon photovoltaic (PV) devices

[2] Osterwald C R, Anevsky S, Barua A K, Chaudhuri P, Dubard J, Emery K, Hansen B, King D, Metzdorf J, Nagamine F, Shimokawa R, Wang Y X, Wittchen T, Zaaiman W, Zastrow A, and Zhang J The World Photovoltaic Scale: An International Reference Cell Calibration Program NREL/CP-510-22942 UC Category: 1250

[3] Inoue T and Yamamura K 1983 Quantum efficiency measurement of a photodiode based on a calorimetric method *J. Appl. Phys.* **54** 6782

[4] Palik E D 1985 Handbook of Optical Constants of Solids (New York: Academic)

[5] CIE 1987 International Lighting Vocabulary *CIE Publication* No. 17.4 (Vienna; CIE)

[6] Saito T 2011 Spectral properties of semiconductor photodiodes *Advances in photodiodes* (Croatia: Intech) chapter 1 pp 3–24

[7] IEC 60904-7:2008 Photovoltaic devices - Part 7: Computation of the spectral mismatch correction for measurements of photovoltaic devices

IOP Conf. Series: Journal of Physics: Conf. Series **972** (2018) 012020 doi:10.1088/1742-6596/972/1/012020

Temperature correction method for commercial CCD array spectrometers used in spectral radiometry measurement

Ling Li, Caihong Dai, Zhifeng Wu and Yanfei Wang

Division of Optics, National Institute of Metrology, Beijing, China

liling@nim.ac.cn

Abstract. Ambient temperature is the main challenge for improving the accuracy of spectral radiometry measurement in the field. In this paper, we present the considerable effect of temperature on the performance of a commercial CCD array spectrometer. Moreover, a temperature correction method for the CCD array spectrometer was developed, which calculated the spectrometer response at each pixel. The temperature correction method is very effective in correcting the spectral response. The deviation between measured and calculated spectrometer responses at the same temperature is less than 1% over the temperature range from 5°C to 40 °C.

1. Introduction

CCD array based spectrometers are widely used in spectral radiance measurements, since they have the key advantage of providing real-time results, compared to traditional mechanical scanning spectrometers. In addition, their small size and light weight nature improves the portability of the instrument. As a result, CCD based array spectrometer is increasingly applied in earth observation [1], remote sensing [2], analytical chemistry [3], photobiological safety [4], and other industries. High accurate calibration and measurement of spectral radiometric are essential for CCD based array spectrometers. With the urgent demand of earth observation and remote sensing, the calibration uncertainties (k=2) of spectral radiometric is required to be less than 2%.

Currently, the laboratory calibrations at National Institute of Metrology P. R. China (NIM) are generally performed at room temperature (25 ±1.0 °C). However, the spectrometers applied in earth observation and remote sensing is usually operated in the field under different ambient temperature conditions. The calibration coefficients determined under room temperature conditions are not applicable to data collected in field situations. This issue has aroused widespread international attention. Starks [5] conducted to characterize temperature sensitivity of a typical silicon-detector-based spectroradiometer, demonstrated the potential errors due to temperature effects, and present a methodology to correct for temperature-induced errors. Salim et al. [6] at National Physical Laboratory (NPL) have studied the nontemperature-stablized effect of a photodiode array spectrometer, and obtained good corrected results with low uncertainty of measurements. Widenhorn [7] investigated the activation energy for the dark current of a back-illuminated CCD changes over the temperature range of 222 to 291 K. Kuusk [8] presented the temperature dependence in the dark output of a spectrometer module at several integration times over a range of ambient temperatures, and developed a hybrid model to represent dark output. Wang [9] indicated that changes in temperature affect the magnitude of both the dark and offset voltages, and the additional signal voltage which is

IOP Conf. Series: Journal of Physics: Conf. Series **972** (2018) 012020 doi:10.1088/1742-6596/972/1/012020

provided by external light is dependent on temperature and integration time together. The total output of the spectrometry system is also affected by the temperature biases of ADC circuitry itself. In addition, the relative position of internal optical elements inside the spectrometer slightly deviates while ambient temperature changes, due to the thermal deformation of the mechanical material [10]. This can affect the accuracy of the measured irradiance. Thus, to achieve more accurate measurement results, it is very important to develop a temperature correction method for CCD array spectrometer used in the field.

2. Experiments

Figure 1 show the schematic of the measurement facility, which is used to evaluate the temperature dependence of the test CCD array spectrometer. It main consists of a standard lamp, thermal chamber, and a commercial fiber optic spectrometer. The instrument is nontemperature-stabilized and containing 2048 pixel elements made of back-thinned CCD detector, that covering the spectral range from 650 to 1100 nm. In order to achieve more similar condition between room measurements and field measurements, thermal chamber is used to change the ambient temperature. The CCD detector is placed in the thermal chamber. The size of operation chamber is 500×400×500mm. The parameters of thermal chamber are -50 °C~100 °C, 20% RH~+100% RH. The spectral accuracy was examined with a mercury argon (Hg-Ar) calibration source (Avantes). The light is guided into the spectrometer through an optical fiber. The aperture is used to minimize the stray light of environment. The spectrometer was changed by temperature controlled chamber and stabilized in setting temperature for 1 hour. The operating temperature of the spectrometer was monitored using a calibrated platinum resistance thermometer (PRT) placed on direct contact with the CCD spectrometer.

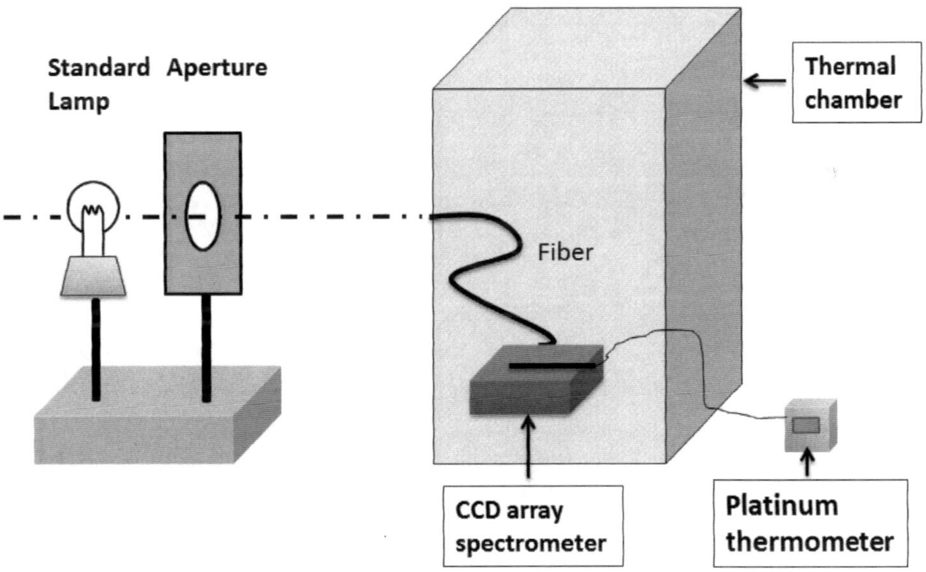

Figure 1. A schematic of the measurement facility.

3. Results

In the wavelength accuracy experiments, it is concluded that no wavelength shift higher than 0.3 nm over the full range of operating temperatures using the spectral lines of the Hg-Ar lamp.

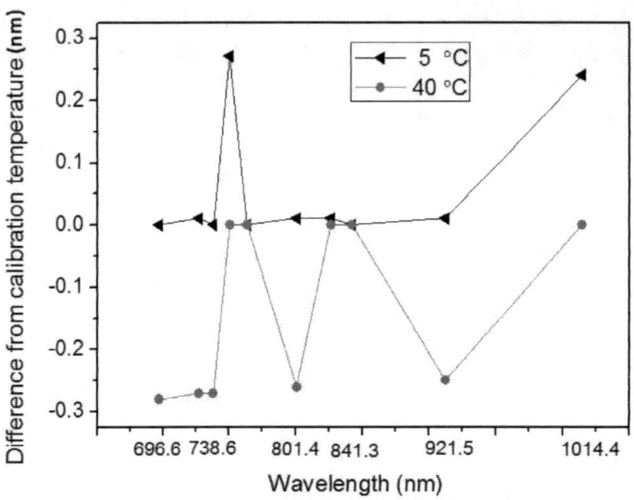

Figure 2. The wavelength difference from calibration temperature (25 °C)

All measurements of the spectrometer use the same integration time. As shown in figure 3, the dark current increased obviously with increasing the temperature. The dark current is the intrinsically noise of CCD. It is due to the thermal excitation of electrons into the conduction band and collection in the CCD wells. The generation of dark electrons is a thermally activated process and as such strongly temperature dependent. The partial potential of CCD is filled with the dark signal, the effective output of the CCD decreases since the capacity of the CCD potential wells is limited [7].

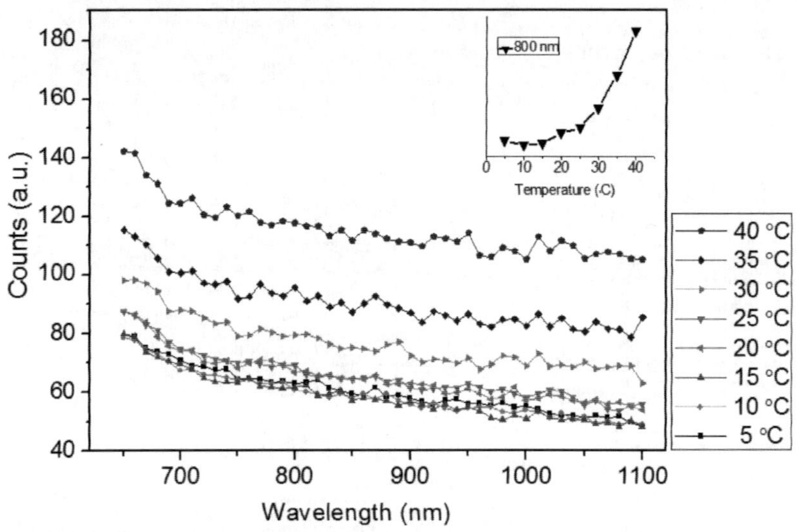

Figure 3. Temperature dependence of dark current at different temperature.

The CCD spectrometer response was monitored over the range from 5 °C to 40 °C using 5 °C temperature intervals. The dark signal was subtracted from the source signal for each measurement. As shown in Figure 4, the CCD spectrometer response changes rapidly with increasing the temperature. When temperature effects were not taken into account, the deviation at the 1060 nm approached ±17% or more, relative to that determined at 25°C. According to Salim et al., this spectral range is close to the band edge of the silicon, which is highly temperature sensitive, and the silicon band edge moved to longer wavelengths as increasing temperature [6].

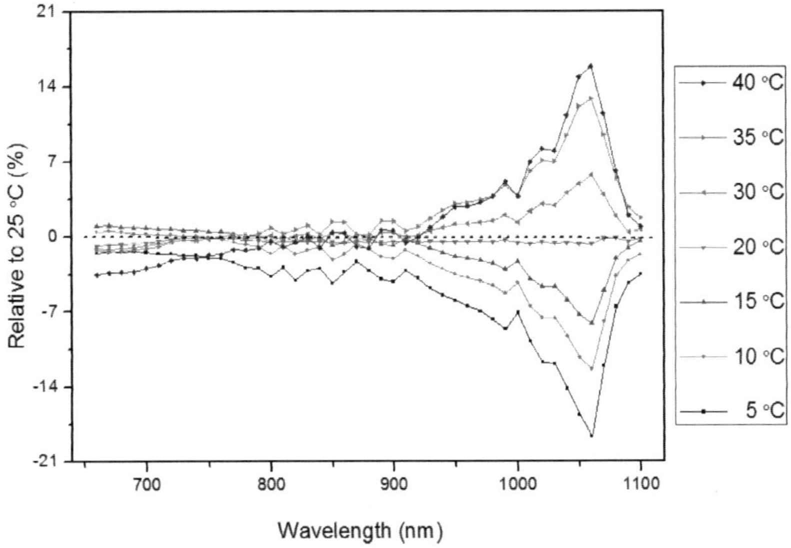

Figure 4. Deviation of the spectrometer response at different temperatures, relative to its response at calibration temperature (25 °C).

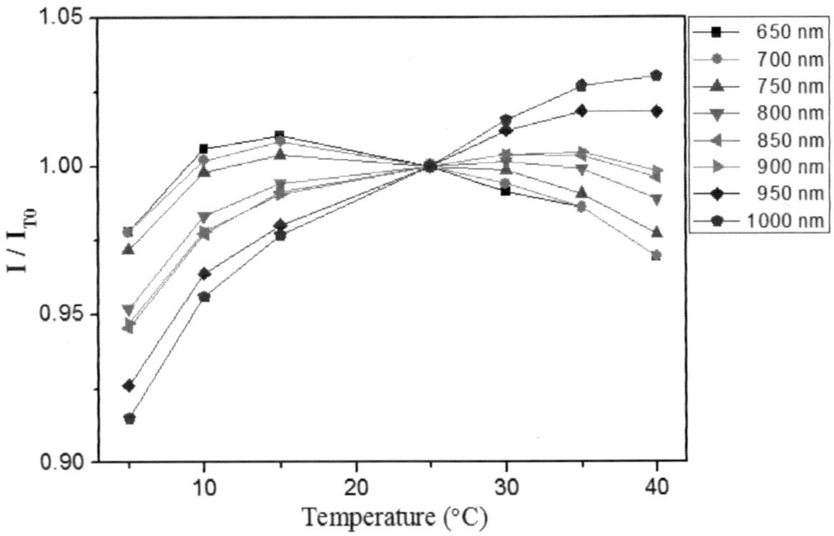

Figure 5. The changes in I_T/I_{T0} under different temperature at different wavelengths.

As shown in figure 5, the change in I_T/I_{T0} under different temperature at any wavelength could be fitted as a polynomial function. The spectral response of CCD spectrometer and the ambient temperature has the following relationship:

$$I(T) = I_{T0} \cdot [a_0 + a_1 T + a_2 T^2 + \cdots + a_k T^k] \quad (1)$$

Where $I(T)$ is the spectral response at experiments temperature T, I_{T0} is the spectral response at calibration temperature T_0. a_0, a_1, a_2,..., a_k is the temperature correction coefficient, which is calculated from I_T/I_{T0}. The estimation of correction coefficients of any ambient temperature can be interpolated from the function. Since the response of each pixel (wavelength) exhibits different behaviour at different ambient temperature, the temperature correction coefficient matrix can be calculated (where n is the pixel; k is the order of fitting).

$$\begin{bmatrix} a_{00} & a_{01} & a_{02} & \cdots & a_{0k} \\ a_{10} & a_{11} & a_{12} & \cdots & a_{1k} \\ \cdots & \cdots & \cdots & \ddots & \cdots \\ a_{n0} & a_{n1} & a_{n2} & \cdots & a_{nk} \end{bmatrix} \tag{2}$$

Equation (1) results in:

$$\begin{bmatrix} I_0(T) \\ I_1(T) \\ \cdots \\ I_n(T) \end{bmatrix} = \begin{bmatrix} I_{0T} & 0 & \cdots & 0 \\ 0 & I_{1T} & \cdots & 0 \\ \cdots & \cdots & \ddots & \cdots \\ 0 & 0 & \cdots & I_{nT} \end{bmatrix} \left(\begin{bmatrix} a_{00} & a_{01} & a_{02} & \cdots & a_{0k} \\ a_{10} & a_{11} & a_{12} & \cdots & a_{1k} \\ \cdots & \cdots & \cdots & \ddots & \cdots \\ a_{n0} & a_{n1} & a_{n2} & \cdots & a_{nk} \end{bmatrix} \begin{bmatrix} 1 \\ T^1 \\ \cdots \\ T^k \end{bmatrix} \right) \tag{3}$$

The equation (3) is a general model, and it can be chosen based on actual experimental results. The model was used to correct the response of spectrometer in the field measurements. According to the experiment results in this paper, the change in I_T/I_{T0} under different temperature (5 °C to 40 °C) at any wavelength could be fitted as a polynomial function of the second order. Any ambient temperature can be interpolated using an interpolation approach between 5 °C and 40 °C. To verify the temperature correction model, the signal of a randomly selected temperature (32 °C) was measured and corrected. The deviation between measured and calculated spectrometer responses at the same temperature is less than 1% (Figure 6). This temperature correction method can be easily used for spectral radiometry measurement in the field.

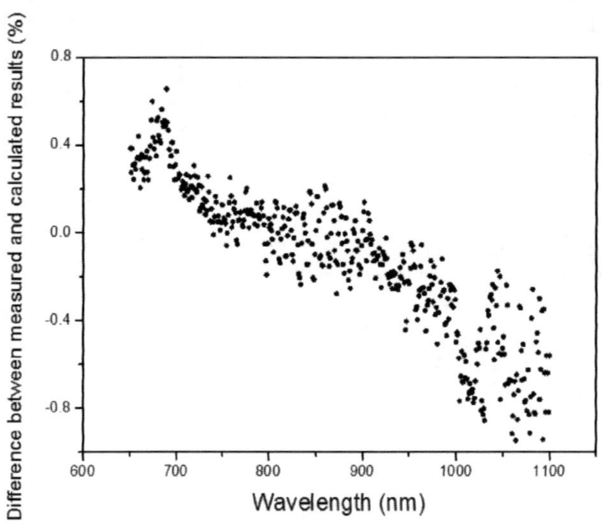

Figure 6. Difference between the measured and calculated results of the CCD spectrometer at randomly selected temperature (32 °C).

4. Conclusion

Temperature dependence is an important issue when the spectrometers used in field applications. This paper demonstrated an experimental approach for assessing and a method for correcting the temperature dependence. The CCD spectrometer show considerable temperature sensitivity, especially for wavelengths longer than 900 nm, while the deviation of wavelength of the nontemperature-stabilized commercial spectrometers is just within ±0.3 nm. The deviation at the 1060 nm approached

IOP Conf. Series: Journal of Physics: Conf. Series **972** (2018) 012020 doi:10.1088/1742-6596/972/1/012020

±17%, relative to that determined at calibration temperature (25 °C). According to the experiment data, the sensitivity of CCD spectrometer has been presented as a function of temperature at each wavelength, and the temperature correction coefficient matrix can be calculated. In addition, the response of spectrometer was corrected method was verified with a randomly selected temperature. The result shows the deviation between measured and calculated spectrometer responses at the same temperature is less than 1%.

Acknowledgments
This work was supported by the National Key R & D Program of China (Grant No. 2016YFF0200304), National Institute of Metrology Fundamental Research Funds (Grant No. 23-AJG1503, No.23-AKY1513).

References

[1] Butler J J and Barnes R A The use of transfer radiometers in validating the visible to shortwave infrared calibrations of radiance sources used by instruments in NASA'S Earth Observing System 2003 *Metrologia* **40** S70-7.
[2] Kieffer H H and Stone T C The spectral irradiance of the moon 2005 *Astron. J.* **129** 2887-2901.
[3] Ojeda C B and Rojas F S Process Analytical Chemistry: Applications of Ultraviolet/Visible Spectrometry in Environmental Analysis: An Overview 2009 *Appl. Spectro. Rev.* **44** 245-65.
[4] Sliney D H Exposure Geometry and spectral environment determine photobiological effects on the human eye 2005 *Photoch. Photobio.* **81** 483-9.
[5] Starks P J, Walter-Shea E A, Schiebe F R and Markham B L Temperature sensitivity characterization of a silicon diode array spectrometer, 1995 *Remote Sens. Environ.* **51** 385-9.
[6] Salim S G R, Fox N P, Theocharous E, Sun T and Grattan K T V Temperature and nonlinearity corrections for a photodiode array spectrometer used in the field 2011 *Appl. Opt.* **50** 866-75.
[7] Kuusk J Dark signal temperature dependence correction method for miniature spectrometer modules 2011 *J. Sens.* **2011** 608157.
[8] Widenhorn R, Blouke M M, Weber A, Rest A and Bodegom E Temperature dependence of dark current in a CCD 2002 *Proc. SPIE.* **4669** 193-201.
[9] Wang H, Song H, Chen Y and Laney S R Correcting temperature dependence in miniature spectrometers used in cold polar environments 2015 *Appl. Opt.* **54** 3162-72.
[10] Wang H, Chen Y, Song H and Laney S R A fiber optic spectrometry system for measuring irradiance distributions in seaice environments 2014 *J. Atmos. Oceanic Tech.* **31** 2844–57.

Liquid nitrogen cryostat for predictable quantum efficient detectors

F Manoocheri[1], T Dönsberg[1], M Sildoja[1,2], M Smíd[3], G Porrovecchio[3] and E Ikonen[1,4]

[1] Metrology Research Institute, Aalto University, Espoo, Finland
[2] Present address: Physikalisch-Technische Bundesanstalt (PTB), Braunschweig, Germany
[3] Czech Metrology Institute (CMI), Praha, Czech Republic
[4] MIKES Metrology, VTT Technical Research Centre of Finland Ltd, Espoo, Finland

E-mail: *farshid.manoocheri@aalto.fi*

Abstract. We present the design and testing of a cryostat to be used with induced junction photodiodes of the Predictable Quantum Efficient Detector (PQED). Long-term reflectance measurements indicate that possible ice growth on the photodiodes at the temperature of liquid nitrogen (LN) is significantly reduced from earlier PQED cryostat designs.

1. Introduction

The predictable quantum efficient detector [1-4] consists of two induced junction photodiodes, which are mounted in a wedged trap configuration to minimise the specular reflectance losses. PQEDs are typically operated at room temperature, where uncertainty of internal quantum deficiency (IQD) of around 100 ppm is achieved. We have also operated PQEDs in liquid nitrogen cryostats to cool the induced junction photodiodes. The cooling reduces dark current, improves the mobility of charge carriers, and therefore reduces the IQD of the photodiodes according to the predictions of photodiode models [5-6]. However, when we cooled the PQEDs in commercially available cryostats, the reflectance of the detector changed temporally [1]. This is due to the growing ice layer on top of silicon photodiodes [7-8].

In this work, we present the design of a new cryostat for operating PQEDs at 79 K temperature. The cryostat is compatible with various versions of the PQED photodiodes. For testing purposes, photodiodes with 310 nm oxide layer from the first production batch of the PQED [9], referred here as round-1, were installed. Long-term reflectance measurements at 79 K indicate that the icing effect is greatly reduced in the new cryostat. In addition, the results of responsivity measurements of round-1 PQED are given. This is useful for the comparison of photodiode responsivity models.

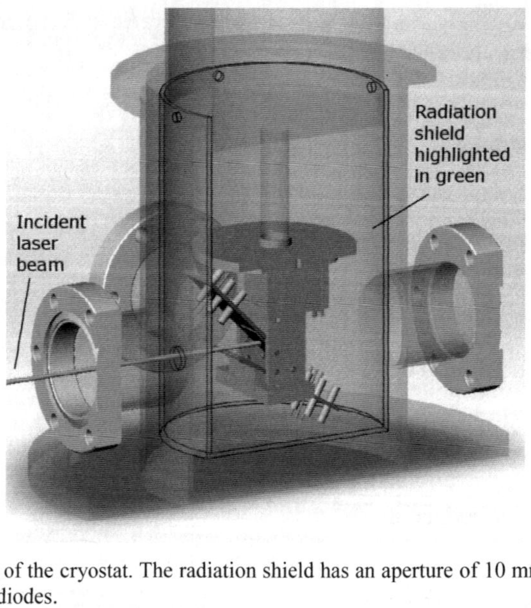

Figure 1. Schematic drawing of the cryostat. The radiation shield has an aperture of 10 mm in diameter for the incident laser beam to reach the photodiodes.

2. Cryostat

The schematic drawing of the cryostat design is shown in figure 1. The temperature of the photodiode holder (the central grey part in figure 1) can be increased and the radiation shield enclosing the photodiodes serves as a cold trap. In addition, the cryostat is equipped with a heatable charcoal getter to trap permeated gases. We pumped the cryostat by using a turbomolecular pump together with a mechanical backing pump. The pumping system is similar as used in the previous publication [1], but the achieved vacuum is 68 μPa, a factor of ten better.

3. Measurement results

Before cooling with liquid nitrogen, the cryostat was baked at 80 °C for 20 hours. Figure 2 shows the reflectance measured over 21 hour time period at LN temperature. Measurements were performed using a stabilized argon ion laser operated at the wavelength of 488 nm. The reflectance has a rate of change around 0.2 ppm/h, which is a factor of five smaller than that previously reported in [1].

Before cooling down the photodiodes, the responsivity of the round-1 PQED was measured at room temperature against a PQED that is similar to those characterized in [2] and [3]. The obtained IQD values for the round-1 photodiodes at room temperature are around 0.2% as shown in table 1. Due to the problems of the packaging and circuitry of chip carriers in the round-1 photodiodes, attempts to determine the IQD values at LN temperatures were not successful at 488 nm wavelength. During comparison measurements against the cryogenic radiometer at the CMI, the IQD of 0.2% at 79 K was obtained for the wavelength of 476 nm. An earlier result was 0.6% for the wavelength of 760 nm in measurements performed at PTB on a similar PQED with one of the photodiodes slightly displaced from its chip carrier [9].

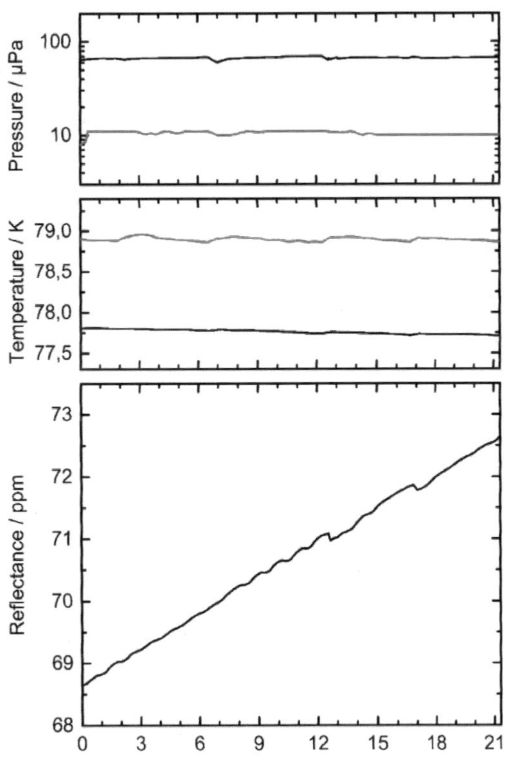

Figure 2. Measured absolute reflectance over 21 hour time period at LN temperature. Pressure was monitored with gauges near the cryostat (black line) and the turbomolecular pump (red line). Platinum sensors were used to measure temperature close to the N_2 vessel (black line) and close to the photodiodes (red line).

Table 1. Measured IQD of the PQED constructed of round-1 photodiodes at various reverse bias voltages and power levels at room temperature. The uncertainties are given at 95% confidence level.

Optical power (µW)	Bias voltage (V)	IQD (%)
100	-5	0.186 ± 0.015
240	-5	0.195 ± 0.021
100	-10	0.195 ± 0.020

4. Conclusions

We have developed a custom designed cryostat for operating the PQEDs at LN temperatures. The device can minimise the growth of ice layers on the photodiodes and therefore achieve a rate of change of reflectance of about 0.2 ppm per hour. This can facilitate the demonstration of the potential near zero IQD with an uncertainty of around 10 ppm predicted for the cooled PQEDs. It is essential that the stability of reflectance is monitored during such measurements of PQEDs at low temperatures.

5. References

[1] Sildoja M *et al* 2013, Predictable quantum efficient detector: I. Photodiodes and predicted responsivity, *Metrologia*, **50**, 385–394.
[2] Müller I *et al* 2013, Predictable quantum efficient detector: II. Characterization and confirmed responsivity, *Metrologia*, **50**, 395–401.
[3] Dönsberg T *et al* 2014, A primary standard of optical power based on induced-junction silicon photodiodes operated at room temperature, *Metrologia*, **51**, 197-202.
[4] Dönsberg T *et al* 2017, Predictable quantum efficient detector based on n-type silicon photodiodes, *Metrologia*, **Open access**, available at http://iopscience.iop.org/article/10.1088/1681-7575/aa85ed
[5] Gran J et al 2012, Simulations of a predictable quantum efficient detector with PC1D, *Metrologia*, **49**, S130–S134.
[6] Tang C K *et al* 2015, Measured and 3D modelled quantum efficiency of an oxide-charge induced junction photodiode at room temperature, 15th International Conference on Numerical Simulation of Optoelectronic Devices (Taipei, Taiwan), available at http://www.nusod.org/2015/nusod15_paper89.pdf
[7] Brown D E *et al* 1996, H2O Condensation coefficient and refractive index for vapor-deposited ice from molecular beam and optical interference measurements, *J. Phys. Chem.*, **100**, 4988-4995.
[8] Berman A 1996, Water vapor in vacuum systems 2011, *Vacuum*, **47**, 327-332.
[9] Müller I *et al* 2011, Predictable Quantum Efficient Detector: II. Characterization Results, Proceedings of NEWRAD 2011 Conference (Hawaii, USA) pp. 209-210.

Acknowledgements
The research leading to these results has received funding from the European Metrology Research Programme (EMRP) project SIB57 "New Primary Standards and Traceability for Radiometry". The EMRP is jointly funded by the EMRP participating countries within EURAMET and the European Union.

IOP Conf. Series: Journal of Physics: Conf. Series **972** (2018) 012022 doi:10.1088/1742-6596/972/1/012022

Stray light correction of array spectroradiometer measurement in ultraviolet

Zhifeng Wu, Caihong Dai, Yanfei Wang and Ling Li

Division of optics, National Institute of Metrology, Beisanhuan Donglu 18, Chaoyang District, Beijing, 100029, China

wuzf@nim.ac.cn

Abstract. For most of the array spectroradiometer, stray light is significant in UV band. Stray light correction of a UV array spectroradiometer is investigated using optical filters. If a group of filters with continuous bandpass are chosen, stray light contribution due to all the bands can be obtained using a numerical algorithm. The array spectroradiometer with the stray light corrected is used to measure the spectral irradiance of several UV lamps. The measurement results are compared to a double monochromator spectroradiometer. When xenon lamp is the array spectroradiometer calibration lamp, after stray light correction, the difference can be improved from nearly 10% to 2.0% in UVC band. When tungsten lamp is the calibration lamp, the difference can be improved from around 90% to less than 20%.

1. Introduction

UV radiometers are used to measure the UV light sources. According to the CIE concept, UV can be divided into UVC (200-280) nm, UVB (280-315) nm and UVA (315-400) nm. However, there is not a one-to-one match between UV radiometers and CIE UV conception. Various radiometers are needed in order to meet the demands of different application fields. For example, narrow band UVC is popular in UV disinfection; the spectral irradiance at 310 nm and 340 nm are needed in aging applications; narrow band UVB is needed in psoriasis treatment; UVA is usable in the crack detection. It's hard to cover all the potential fields just using radiometers, especially with new growth in UV demand. On the other hand, UV radiometers are often used to measure UV light sources different from those used by the manufacturer. The spectral mismatch can bring big problem for the measurement.

It's obvious that traditional scanning spectroradiometer can solve the problem. However, it is large and not convenient to take along. In addition, the scanning speed also limits its use. Since compact array spectroradiometer has a fast acquisition speed and is portable, it is widely used in photometric and radiometric measurement in recent years. The characteristic of the array spectroradiometer was investigated during the past ten years, such as the bandwidth, the stability and the stray light [1-9]. However, report on the UV spectroradiometer application in direct use is not much due to the significant stray light in UV band. Zong has proposed a stray light correction method using tunable lasers [1]. Feldman has implemented a digital micromirror device in array spectroradiometer to reduce the stray light [6]. Shaw has investigated the stray light of array spectrometers using a series of cut-on filters [7]. Compared to the tunable laser method, the filter method is much cheaper and can save a lot of time. Shindo has investigated the stray light of array-based spectrometers in UV band using cut-on filter correction method [8]. Kenji analysed the heterochromatic stray light and proposed a practical method to correct the stray light [9]. In this paper, optical filters method is also used to analyze the stray light effect in UV region.

2. Experiment and results

2.1. Array spectroradiometer

The optical setup of array spectroradiometer is shown in Figure 1. Through the entrance slit, the light is collimated by the collimating mirror and illuminates the grating. The grating diffracts the light spectrum and makes different wavelengths received by different parts of the CCD detector. Each pixel of the CCD detector and the corresponding wavelength has a one-to-one relationship, which can be defined by the atomic lines of spectral calibration lamps, such as Hg-Ar lamp. UV array spectroradiometer covering 200 nm to 460 nm is investigated, with 2048 pixels and 1800 lines/mm grating. Since array spectroradiometer has 2048 pixels, the wavelength interval between each pixel is about 0.127 nm. The wavelength accuracy is found to be less than 0.1 nm.

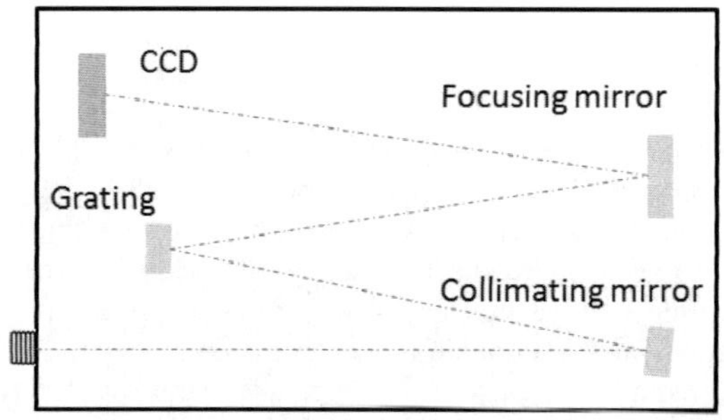

Figure 1. Optical setup of array spectroradiometer.

2.2. Stray light test of array spectroradiometer using optical filter

If a single wavelength laser is used to irradiate the array spectroradiometer, the response of the array spectroradiometer at other wavelength is due to stray light. Similarly, if a lamp with bandpass optical filter is used, the response of the array spectroradiometer out of filter bandpass region should be due to stray light.

A Schott GG450 filter is mounted in front of the array spectroradiometer. Figure 2 shows the stray light ratio in UV band. The blue and green lines are the ratios with and without filter using xenon lamp and tungsten lamp, while the red line is the transmittance of the optical filter. It can be seen that the optical filter has a transmittance less than 0.1% in 200 nm – 400 nm. However, the ratios are much larger than the filter transmittance, especially when tungsten lamp is used. The difference between the ratios and the transmittance is the stray light, which is due to the visible and infrared spectra of the lamps. The stray light dominates when tungsten lamp is used, while the stray light accounts for less than 6.0% for the xenon lamp.

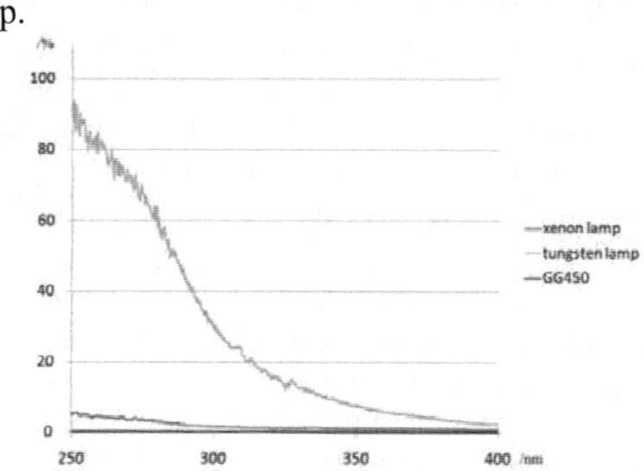

Figure 2. The UV stray light ratios using xenon lamp and tungsten lamp.

Figure 3 shows the relative spectral power distribution of xenon lamp and tungsten lamp at the same distance. The ratio of spectral irradiance at 310 nm and 253.7 nm to the integral spectral irradiance (500-1100) nm is about 2.5×10^{-5} and 2.3×10^{-6} for tungsten lamp, while the ratio is 5.6×10^{-4} and 2.4×10^{-4} for xenon lamp. It's obviously that xenon lamp has a much large UV spectral proportion and the stray light ratio is much smaller.

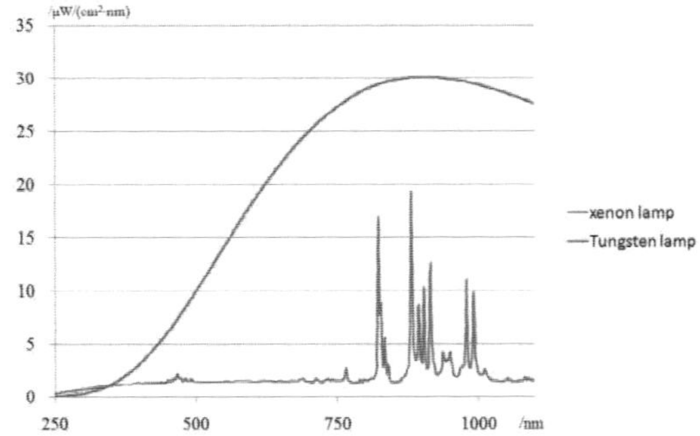

Figure 3. The spectral power distribution of xenon lamp and tungsten lamp.

2.3. *Stray light correction method*

Since Schott GG450 filter has a high transmittance in visible and infrared, the stray light ratio due to visible and infrared can be measured. If UVA, UVB and UVC bandpass filters are used, the stray light ratio due to UVA, UVB and UVC band can be obtained. Equation (1) and (2) describe the relationship between the real spectral irradiance and the measured spectral irradiance when an optical filter is used. $E_{lamp}(\lambda_0)$ is the spectral irradiance of the lamp at wavelength λ_0, while $E'_{measure}(\lambda_0)$ is the spectral irradiance measured by the array spectroradiometer. $E_{lamp}(\lambda)$ is the spectral irradiance of the lamp at wavelength λ, $T(\lambda)$ is the transmittance of the filter at wavelength λ, $R(\lambda\rightarrow\lambda_0)$ denotes the stray light contribution at wavelength λ_0 due to wavelength λ. $S_{with}(\lambda_0)$ is the spectroradiometer signal obtained with the filter, $S_{without}(\lambda_0)$ is the signal obtained without the filter, λ_i and λ_f are the lower and upper wavelengths in summation. If no filter is used, $T(\lambda)=1$. Then the right term of equation (1) will be $S_{without}$. It's easy to see that the measured spectral irradiance is bigger than the real spectral irradiance of the lamp due to the stray light. As an approximation, $R(\lambda\rightarrow\lambda_0)$ in equation (2) is moved out of the summation, $T(\bar{\lambda})$ is average transmittance between λ_i and λ_f.

$$E_{lamp}(\lambda_0)T(\lambda_0)+\sum_{\lambda_i=200(\lambda\neq\lambda_0)}^{\lambda_f=280}E_{lamp}(\lambda)T(\lambda)R(\lambda\rightarrow\lambda_0)+\sum_{\lambda_i=280(\lambda\neq\lambda_0)}^{\lambda_f=320}E_{lamp}(\lambda)T(\lambda)R(\lambda\rightarrow\lambda_0)+\sum_{\lambda_i=320(\lambda\neq\lambda_0)}^{\lambda_f=400}E_{lamp}(\lambda)T(\lambda)R(\lambda\rightarrow\lambda_0)+\sum_{\lambda_i=400(\lambda\neq\lambda_0)}^{\lambda_f=1100}E_{lamp}(\lambda)T(\lambda)R(\lambda\rightarrow\lambda_0)=E'_{measure}(\lambda_0) \quad (1)$$

$$R(\lambda\rightarrow\lambda_0)\sum_{\lambda=\lambda_i}^{\lambda=\lambda_f}E_{lamp}(\lambda)=\frac{S_{with}(\lambda_0)/S_{without}(\lambda_0)/T(\bar{\lambda})}{1-S_{with}(\lambda_0)/S_{without}(\lambda_0)} \quad (2)$$

A group of filters consisting of UVC, UVB, UVA and Schott GG400 filter can cover all the wavelengths from 200 nm to 1100 nm. The transmittance of the UVA, UVB and UVC filters are listed in figure 4. With the four filters, the R parameter of a certain lamp can be obtained. According to equation (1), the relationship of the real spectral irradiance of the lamp and the measured spectral irradiance can be described using equation (3).

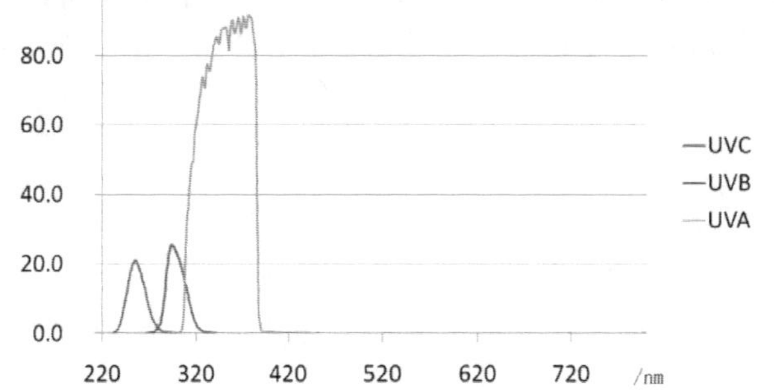

Figure 4. The transmittance of the filters.

$$\begin{bmatrix} 1 & R_{\text{uvb}\to\text{uvc}} & R_{\text{uva}\to\text{uvc}} & R_{\text{uvo}\to\text{uvc}} \\ R_{\text{uvc}\to\text{uvb}} & 1 & R_{\text{uva}\to\text{uvb}} & R_{\text{uvo}\to\text{uvb}} \\ R_{\text{uvc}\to\text{uva}} & R_{\text{uvb}\to\text{uva}} & 1 & R_{\text{uvo}\to\text{uva}} \\ R_{\text{uvc}\to\text{nonuv}} & R_{\text{uvb}\to\text{nonuv}} & R_{\text{uva}\to\text{nonuv}} & 1 \end{bmatrix} \begin{bmatrix} E_{\text{uvc}} \\ E_{\text{uvb}} \\ E_{\text{uva}} \\ E_{\text{non-uv}} \end{bmatrix} = \begin{bmatrix} E_{\text{uvc}}'' \\ E_{\text{uvb}}'' \\ E_{\text{uva}}'' \\ E_{\text{non-uv}}'' \end{bmatrix} \quad (3)$$

In equation (3), the first column is a matrix. Subscript uvc, uvb, uva and uvo correspond to UVC band, UVB band, UVA band and out of UV band. The second column and third column correspond to the real and measured spectral irradiance of the light source in UVC, UVB, UVA and out of UV band. For example, if $\lambda_0 \in UVC$ and UVB filter is used, the right term of equation (2) is $R_{\text{uvc}\to\text{uvb}}$, which is stray light contribution in UVC band due to UVB band. As a simplification, $R_{\text{uvc}\to\text{uvc}(\lambda\neq\lambda_0)}$ is omitted. Since the R parameters in the first column can be obtained with the help of the four filters, the stray light can be easily corrected. After transformation, the real spectral irradiance of the lamp in the ultraviolet band can be obtained.

2.4. Stray light correction results

In order to test the capability of the stray light correction method, comparison was carried out in contrast with a double monochromator spectroradiometer. The comparison is done according to the following procedure. First, xenon lamp with known spectral irradiance is used to calibrate the array spectroradiometer and the double monochromator spectroradiometer. Second, the stray light ratio is test using four filters and the xenon lamp. Third, the stray light of the array spectroradiometer is corrected. Finally, the array spectroradiometer and the double monochromator spectroradiometer are used to measure the same UV lamps, and the measurement results are compared.

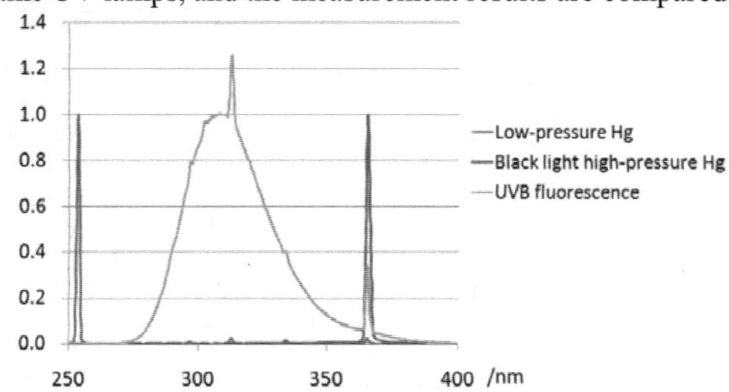

Figure 5. The spectra of UV lamps.

IOP Conf. Series: Journal of Physics: Conf. Series **972** (2018) 012022 doi:10.1088/1742-6596/972/1/012022

The spectra of the UV lamps are shown in figure 5. Black light high-pressure Hg lamp and low-pressure Hg lamp are both narrow band lamps, while UVB fluorescence is a broad band lamp. Table 1 shows the comparison results using xenon lamp. For the UVA, UVB, and UVC lamp, integral spectral irradiance of (360-370) nm, (280-320) nm, and (250-260) nm are compared. The result in the second column using double monochromator spectroradiometer is set to 1.000. The third and fourth columns show the ratio using array spectroradiometer before and after stray light correction. It can be seen that after stray light correction, the difference in UVC band can be improved from 9.4% to 2.0%. In addition, the difference in UVB and UVA can be improved to less than 0.5%. Table 2 shows the comparison results using tungsten lamp calibration. For the UVA and UVB lamp, the difference can be improved to less than 5.0%, which implies the feasibility using filter correction method. However, the stray light level is too high in UVC band, where the parameters R in equation (2) are much greater than 1, while the parameters R are less than 0.1 when using the xenon lamp. Before correction, the difference in UVC band is greater than 90%. The stray light is nearly 10 times bigger than the actual signal. After stray light correction, the difference is less than 20%. In figure 2, oscillation can be seen which means the signal-to-noise is not good. Considering the bad signal-to-noise and approximation of the method, the UVC correction result is acceptable.

Table 1. Comparison using xenon lamp calibration.

Lamp	Double monochromator	Array spectroradiometer	After stray light correction
Low-pressure Hg	1.000	0.906	0.980
UVB fluorescence	1.000	0.982	0.997
Black light high-pressure Hg	1.000	0.992	1.001

Table 2. Comparison using tungsten lamp calibration.

Lamp	Double monochromator	Array spectroradiometer	After stray light correction
Low-pressure Hg	1.000	0.093	0.816
UVB fluorescence	1.000	0.781	1.049
Black light high-pressure Hg	1.000	0.925	0.983

3. Conclusion

For most of ultraviolet array spectroradiometer, the quantum efficiency of CCD detector in visible and infrared band differs not too much from the ultraviolet band. Lamp with large spectral power distribution variation from UV to other wavelength will bring significant stray light in the UV band. With a group of optical filters, the stray light due to different wavelength range can be corrected. After stray light correction, consistency can be seen especially for the xenon lamp.

Acknowledgements

This work was supported by the National Institute of Metrology Fundamental Research Funds under grant No.23-AKY1513, and National Natural Science Foundation of China under grant No.41305140.

References

[1] Yuqin Zong, Steven W. Brown, B. Carol Johnson, Keith R. Lykke and Yoshi Ohno 2006

Simple spectral stray light correction method for array spectroradiometers *Appl. Opt.* **45** 1111-19

[2] Saulius Nevas, Annette Teuber, Armin Sperling and Matthias Lindemann 2012 Stability of array spectroradiometers and their suitability for absolute calibrations *Metrologia* **49** S48-52.

[3] Saber G. R. Salim, Nigel P. Fox, Evangelox Theocharous, Tong Sun and Kenneth T. V. Grattan 2011 Temperature and nonlinearity corrections for a photodiode array spectrometer used in the field *Appl. Opt.* **50** 866-75

[4] Emma R. Woolliams, R´ejean Baribeau, Agnieszka Bialek and Maurice G Cox 2011 Spectrometer bandwidth correction for generalized bandwidth function *Metrologia* **48** 164-72

[5] Zhifeng Wu, Caihong Dai, Yanfei Wang and Juhong Zou 2016 Measurement of optical parameters for array spectroradiometers *Opt. Precision Eng.* **24** 77-82

[6] A Feldman, T Burnitt, G Porrovecchio, Smid M, Egli L, J Grober and K M Nield 2014 Diode-array UV solar spectroradiometer implementing a digital micromirror device *Metrologia* **51** S289-92

[7] Michael Shaw and Terasa Goodman 2008 Array-based goniospectroradiometer for measurement of spectral radiant intensity and spectral total flux of light sources *Appl. Opt.* **47** 2637-47

[8] Francois Shindo, Emma Woolliams, Barry Scott and Subrena Harris 2013 Characterization and calibration of compact array spectrometers in the ultraviolet spectral region *AIP Conf. Proc.* **1531** 809-12

[9] Kenji Imura, Yuzuru Yamamoto, Yohsuke Takebe and Yoshiroh Nagai 2017 Practical method for correcting heterochromatic stray light in dual-channel spectrographs for colorimetry *Color Res. Appl.* **42** 431–39

IOP Conf. Series: Journal of Physics: Conf. Series **972** (2018) 012023 doi:10.1088/1742-6596/972/1/012023

APMP Pilot Study on Transmittance Haze

[1]Wen-Chun Liu, [2]Jisoo Hwang, [3]Annette Koo, [4] Houping Wu, [5]Rojana Leecharoen, [6]Hsueh-Ling Yu

[1]Center for Measurement Standards, ITRI, Hsinchu, Taiwan
[2]Korea Research Institute of Standards and Science, Daejeon, Korea
[3]Measurement Standards Laboratory, Lower Hutt, New Zealand
[4]National Institute of Metrology, Beijing, China
[5]National Institute of Metrology (Thailand), Pathumthani, Thailand
[6]Singularity Optics Corp., Hsinchu, Taiwan

Corresponding e-mail address: liuwenchun@itri.org.tw

Abstract. Five NMIs within APMP, including CMS/ITRI, MSL, NIM, NIMT and KRISS from TCPR applied to the APMP technical committee initiative project for funding to carry out a pilot comparison of transmittance haze in 2012. The project started in 2014 and the final report was completed at the end of 2016. In this pilot comparison, three different haze standards were adopted, and transmittance haze for each standard was measured according to ASTM D1003 or ISO 14782. This paper presents the first results of an APMP pilot study of transmittance haze and the analysis of the variation among different haze measurement systems which are commonly used. The study shows that the variables such as sphere multiplier, transmittance distribution, fluorescence of samples and optical path of the incident beam cause discrepancies among NMIs and highlight deficiencies in current documentary standards.

1. Introduction

The definition of transmittance haze (*TH*) is the ratio of the diffuse transmittance (*DT*) to the total transmittance (*TT*). Transmittance haze is usually applied to characterize the light scattering caused by particles in transparent or translucent materials such as plastic, glass, and liquid, or by their rough surfaces. The importance of haze measurement has increased recently due to the development of optoelectronic industries, such as liquid crystal displays, thin film solar cells, touch panels, lampshades, and so on. The intense competition in these developing industries makes quality control for materials more and more important. Haze and luminous transmittance data are especially useful for quality control as they can be used to identify internal defects of the material. Therefore, much attention is being paid to haze measurement accuracy and traceability in recent years. In response to the increasing demands from industries, establishing the standard for transmittance haze and ensuring its global equivalence are the responsibilities of NMIs.

To satisfy the needs of industries, several national metrology institutes (NMIs) within the Asia Pacific Metrology Programme (APMP), including CMS/ITRI, MSL, NIM, NIMT and KRISS from the technical committee for photometry and radiometry (TCPR) applied to the APMP technical committee initiative project for funding to carry out a pilot comparison of transmittance haze in 2012, and CMS/ITRI was the project overseer. These experiments for the comparison were performed between 2014 and 2015, and data analysis and final report were completed in 2016.

Three kinds of haze standard from different manufacturers were adopted, and *TH* for each standard was measured according to ASTM D1003 [1] or ISO 14782 [2]. CMS/ITRI measured *TH* of all samples by ASTM D1003 and ISO 14782, and then sent them to participant NMIs. All participants measured their artefacts by ASTM D1003, and MSL also measured haze by ISO 14782.

To identify the likely causes of discrepancies in the comparison, some other measurement methods such as Bidirectional Transmittance Distribution Function (BTDF), a newly developed double-compensation method developed by CMS/ITRI [3,4], or double beam method [4], were also used in this study. The combination of methods and NMIs was structured to allow cross-checking of different methods. The results of the pilot comparison are shown in this report, and the causes of the measurement differences between NMIs and CMS/ITRI are also discussed.

2. Method of the Comparison

Five sets of haze plates were used as the artefacts for this comparison; each set has three plates with nominal transmittance haze values of close to 20 % from three manufacturers: BYK Gardner, Nippon Denshoku, and Diffusion Systems. The measurement was under illuminants C or D65 according to ASTM D1003 or ISO 14782, the methods of which are shown schematically in Figure 1(a) and 1(b), respectively. Compared to the structure of ASTM D1003, a compensative port is added on the integrating sphere in ISO 14782. Supposing that the light trap could capture the light completely, the ratio (I_4/I_2) would be the main contribution to *TH*. The components on the sphere in procedure (ii) and (iv) of ISO 14782 in Figure 1(b) are the same, and thus there is no non-equivalence problem of the integrating sphere between the two procedures (in contrast to comparing (ii) and (iv) of Figure 1(a)). The added compensative port makes accurate haze measurement by ISO 14782 possible.

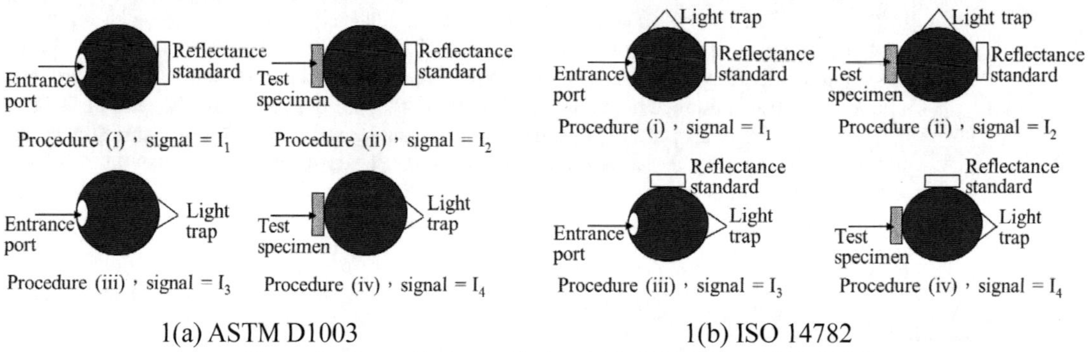

Figure 1. The method specified in ASTM D 1003 and ISO 14782.

$TT = I_2/I_1$, $DT = [I_4 - I_3(I_2/I_1)]/I_1$, $TH = I_4/I_2 - I_3/I_1$

NIMT uses a commercial double-beam spectrophotometer with a diffuse reflectance accessory to measure *TH*. In such an instrument, the haze artefact is put on the entrance port of the sphere, and a white reference or light trap is placed on the port originally used for the tested reflectance sample. The measurement process is therefore like ASTM D1003 shown in Figure 1(a). However, an additional reference port on the sphere for a reference beam is available to monitor the stability of beam and to correct for the substitution error due to the presence of the sample surface. Thus, its signal I_x at each step is corrected automatically by the reference beam according to equation (1) where $I_{s,x}$ is the signal obtained at procedure x and $I_{r,x}$ is obtained when the reference beam enters the sphere. Since there is no documentary standard for using double beam system to measure *TH*, in this pilot study, *TH* obtained from such an instrument is still considered as using ASTM D1003.

$$I_x = I_{s,x}/I_{r,x} \quad (1)$$

At KRISS, the artefacts were measured under illuminant D65 by ASTM D1003 against the standards traced to NPL in accordance with BS:2782 Part 5[5], shown in Figure 2, while other participants measured haze under illuminant C according to the definition of transmittance haze shown in Figure 1. The pilot calculated the difference between KRISS and the pilot under illuminant D65.

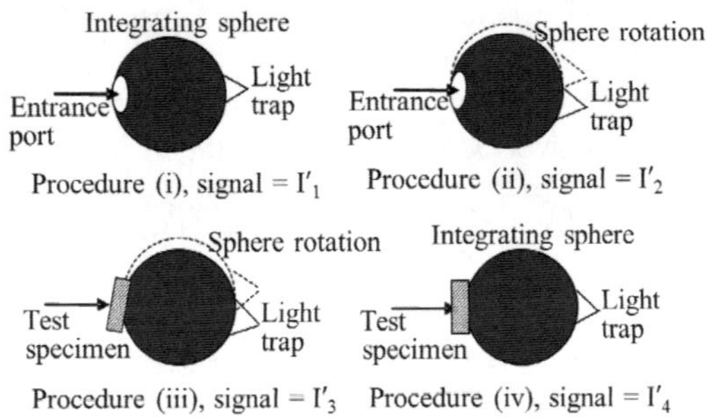

Figure 2 The transmittance haze procedure by BS:2782 Part5.
$TT = I'_3/I'_2$, $DT = [I'_4 - I'_1(I'_3/I'_2)]/I'_2$, $TH = I'_4/I'_3 - I'_1/I'_2$

3. Results and Discussion

Five sets of artefacts were prepared and first measured by CMS/ITRI, and then each participant NMI received its own artefacts. After all the measurements were finished, the data was analyzed by CMS/ITRI. The results and discussion of ASTM D1003 and ISO 14782 follow.

3.1 ASTM D1003

The haze results by ASTM D1003 are shown in Table 1, where the U_{NMI} is the expanded *TH* uncertainty of participant NMIs. As all participants' results are compared with the pilot, the difference (δ) was calculated by equation (2), where S_{NMI} is the measurement result from the participant, and S_{CMS} is the measurement result from CMS/ITRI.

$$\delta = S_{NMI} - S_{CMS} \quad (2)$$

Table 1 The differences between NMIs and the pilot by ASTM D1003 under illuminant C

Manufacturer	NMI	Difference (*TH*)	Difference (*DT*)	Difference (*TT*)	U_{NMI} ($k = 2$)
BYK Gardner	NIM	1.13 %	0.92 %	-0.61 %	0.48 %
	NIMT	0.10 %	-0.04 %	-0.64 %	0.19 %
	MSL	-3.40 %	-3.06 %	0.37 %	0.26 %
	KRISS*	2.28 %	1.91 %	-0.76 %	0.6 %
	CMS/ITRI	0.00 %	0.00 %	0.00 %	0.40 %
Nippon Denshoku	NIM	0.69 %	0.57 %	-0.14 %	0.42 %
	NIMT	-0.14 %	-0.44 %	-1.86 %	0.19 %
	MSL	-3.51 %	-2.97 %	0.26 %	0.54 %
	CMS/ITRI	0.00 %	0.00 %	0.00 %	0.40 %
Diffusion Systems	NIM	0.32 %	0.22 %	-0.31 %	0.40 %
	NIMT	0.09 %	-0.17 %	-1.50 %	0.19 %
	MSL	-2.80 %	-2.22 %	1.40 %	0.38 %
	KRISS*	2.26 %	1.92 %	-0.25 %	0.5 %
	CMS/ITRI	0.00 %	0.00 %	0.00 %	0.40 %

*KRISS measured haze under illuminant D65, and its *DT* was calculated according to the definition ($DT = TT \times TH$).

The trends in all kinds of artefacts are very similar. The largest *TH* and *DT* were measured at KRISS, the smallest were at MSL, and the closest were between NIMT and CMS/ITRI.

The main reason for the haze variation among the NMIs is the non-equivalent sphere reflectance in procedure (ii) and (iv) of ASTM D1003 in Figure 1(a) as the final estimate of *TH* is proportional to the sphere multiplier ratio (*R*) between these two configurations. The sphere multiplier (*M*) for each measurement step is shown in equation (3) where ρ_o is the initial reflectance for incident flux, ρ_s is the reflectance for integrating sphere wall, ρ_i is the reflectance for the port and f_i is the fractional area of the port to the sphere internal surface area [6].

$$M = \frac{\rho_o}{1 - \rho_s\left(1 - \sum_{i=1}^{n} f_i\right) - \sum_{i=0}^{n} \rho_i f_i} \quad (3)$$

The ratio of procedure (iv) to (ii), $R = M_{(iv)}/M_{(ii)}$, for MSL, NIM and CMS/ITRI is 0.75, 0.98 and 0.95, respectively. At KRISS, the reference standards were calibrated by BS:2782 Part 5 whose structures of the sphere are the same in procedures (iii) and (iv), and also the same in procedures (i) and (ii), so R_{KRISS} is 1. At NIMT, the non-equivalence of sphere structures was corrected by reference beam shown in equation (1), so R_{NIMT} is also 1. The difference of participants' measurements from CMS/ITRI measurements are shown in Figure 3 for the original measurements made according to ASTM, and for measurements corrected by the ratios indicated above according to equation (4).

$$\delta = S_{\text{NMI}}/R_{\text{NMI}} - S_{\text{CMS}}/R_{\text{CMS}} \tag{4}$$

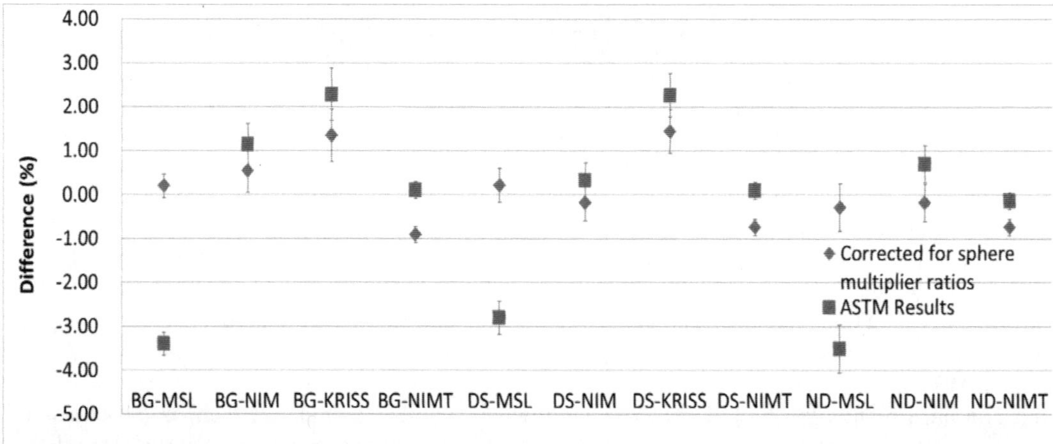

Figure 3 The difference between NMIs and the pilot before and after correction. Uncertainties are shown at a $k = 2$ level. (BG = BYK Gardner, DS = Diffusion Systems, ND = Nippon Denshoku)

Through the results in Figure 3, it can be noticed that except for the data from NIMT, all differences became close to zero after being corrected by the multiplier factor. At MSL, the three differences with the expanded uncertainties cover zero after corrections, and at NIM, it is only a little over zero for the BYK Gardner sample. The sphere multiplier is the key factor for the deviation for these two NMIs. The largest variation from the BYK Gardner sample at NIM might be due to fluorescence. The BYK Gardner sample was fluorescent when it was illuminated by a UV lamp, but the other two kinds of the haze samples had no such characteristics.

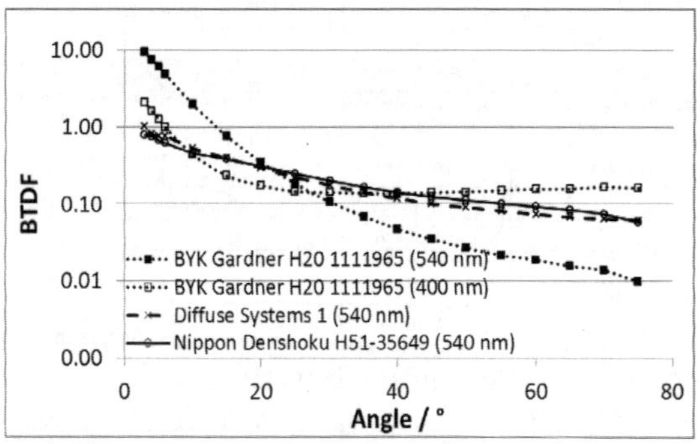

Figure 4(a) BTDF at 540 nm for the three samples at 0° angle of incidence (and at 400nm for BYK sample

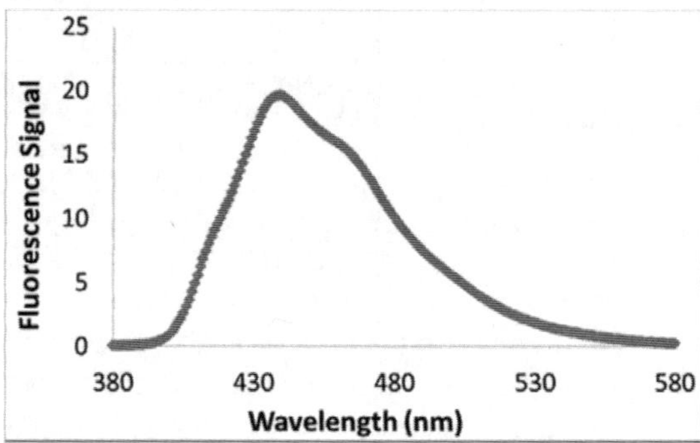

Figure 4(b) The fluorescence emission spectrum of the BYK Gardner sample from 380 nm to 580 nm (an excitation wavelength of 300 nm).

Measurements of the BTDF for the three samples were made on the MSL goniospectrophotometer at a 0° angle of incidence. The BTDF for the samples is shown in Figure 4(a) at 540 nm. These curves are representative of the BTDF at all other wavelengths for each sample, except for the shortest wavelengths for the BYK Gardner sample (also shown at 400 nm in Figure 4(a)). In fact, it also appears that the light scattered to large angles is greater for the BYK sample at 400 nm, than for either of the other samples. This is likely to be the isotropic fluorescence. To confirm if the BYK Gardner artefact is fluorescent, its fluorescence emission spectrum was measured in a commercial fluorometer shown in Figure 4(b). At MSL, monochromatic light is incident on the haze sample, and thus the fluorescence was observed at short wavelengths, even though the fluorescence is emitted at visible wavelengths. In such the measurement system, the fluorescence has minimal effect on *TH* because V(λ) is very small at short wavelengths. At NIM, a halogen lamp and a suitable filter were used to form a light source close to illuminant C, and the transmitted light and induced fluorescence were detected together by a V(λ) detector. At CMS/ITRI, broadband light beam from a halogen lamp is incident on the sample and then the signals were measured by a spectrometer. Thus the fluorescence at the two NMIs was measured at correct wavelengths. Compared to the light source at CMS/ITRI close to illuminant A, the one at NIM to simulate illuminant C is stronger at short wavelengths, so the fluorescence was more pronounced at NIM.

The persistent differences of KRISS can be attributed to a non-normal incident angle in procedure (iii) of BS:2782 Part 5. The effective illuminance on the surface should be multiplied by a cosine factor since the incident angle is not 0°. Thus, I_2 is larger than I'_3 compared to the relative intensity in ASTM and BS methods. In addition, a light trap on the integrating sphere may capture some light, and that would also lower I'_3. Therefore, the *TH* mainly contributed to by the ratio (I'_4/I'_3) is larger.

At NIMT, the non-equivalence of sphere structures was corrected by a reference beam so its data should be close to those of ISO 14782. The original data at NIMT agreeing with CMS/ITRI can be attributed to an incident angle of 8°, not zero. To check the effect of the incident angle, CMS/ITRI measured *TH* of its own three samples by using 0/d (actual incident angle: 3°20') and 8/d integrating sphere accessories installed in the commercial spectrophotometer and by ISO 14782, and the results are shown in Table 2. The data from the 0/d accessory is near the data obtained by ISO 14782 as expected. From the ratio of *TH* using each accessory, the effect of the incident angle is largest for the BYK Gardner sample, and is similar for the other two samples. It is inferred that the light beam after passing through BYK Gardner sample has a higher portion of diffuse transmittance influenced by the incident angle. Compared to the other two samples, a greater fraction of the light scattered by the BYK Gardner sample is within ~ 20° of normal (shown in Figure 4(a)), and thus BYK Gardner sample is more severely affected when the flux is weakened due to a non-perpendicular light beam on the surface of the sample.

Table 2 Haze results by ISO 14782 and the 0/d and 8/d accessories

TH	ISO 14782	0/d Accessory	8/d Accessory	Ratio($H_{0/d}$ / $H_{8/d}$)
Diffusion Systems	16.82 %	16.76 %	16.19 %	1.04
Nippon Denshoku	18.59 %	18.52 %	17.69 %	1.05
BYK Gardner	18.01 %	18.22 %	16.51 %	1.10

3.2 ISO 14782

Except for the pilot, only MSL also measured the artefacts by ISO 14782. The results are shown in Table 3, and "Reference" means the data of CMS/ITRI and its uncertainty. Compared to the difference of *TH* and *DT* between MSL and CMS in ASTM D1003, the difference by ISO 14782 became significantly smaller, and the *TH* from all kinds of sample agree within the uncertainties. The *TH* can be measured accurately by ISO 14782 as the sphere effect is corrected in this method.

Table 3 Results by ISO 14782 under illuminant C

Manufacturer	Difference (*TH*)	Difference (*DT*)	Difference (*TT*)	U_{NMI} (k = 2)
BYK Gardner	0.23 %	0.32 %	0.56 %	0.34 %
Nippon Denshoku	-0.73 %	-0.59 %	0.92 %	0.70 %
Diffusion Systems	-0.32 %	-0.12 %	0.21 %	0.48 %
Reference	0.00 %	0.00 %	0.00 %	0.40 %

In the results of ASTM D1003 and ISO 14782, the largest deviation occurred in the sample of Nippon Denshoku. Among the three kinds of artefacts, only Diffusion Systems sample has no frame. Neither the BYK Gardner sample nor the Nippon Denshoku sample could be presented directly to the sphere port, but the larger deviation was in the Nippon Denshoku sample. To examine the recess effect on the haze, some tests on all three samples were carried out at MSL. Figure 5 shows that the haze changes little when BYK Gardner sample was recessed from the port. The difference between the BYK Gardner sample and the other two samples is consistent with the difference seen in their

IOP Conf. Series: Journal of Physics: Conf. Series **972** (2018) 012023 doi:10.1088/1742-6596/972/1/012023

transmittance distribution (shown in Figure 4(a)). Light scattered to very wide angles 'should' be included in the haze estimate. In the two samples, the effect of recess on Nippon Denshoku sample is severe, and that makes the largest deviation in the Nippon Denshoku sample.

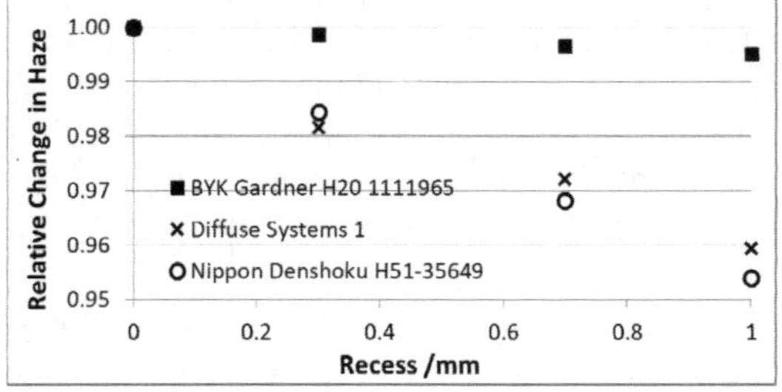

Figure 5 Effect of recess of the sample from the port on the measured value of *TH*

4. Conclusion

The haze measured by ASTM D1003 was highly dependent on the sphere multiplier because of the different configuration of the integrating sphere ports in each procedure of the measurement process. Most of the variation between participants can be accounted for by the sphere multiplier.

Among the three different artefacts, two of them had a frame that may prevent the sample from touching the port of the sphere directly and influences the fraction of light scattered into the sphere. The recess of the sample from the port makes *TH* decrease more seriously when samples are more highly scattering at moderate angles from normal. The transmittance distribution of the sample also influences the measurement results at different incident angles. When the incident angle is not 0°, *TH* of the sample decreases more if most of its scattering is within a smaller angle.

One of the samples exhibited inducement in the UV region, so the UV portion of the incident light might determine the relative intensity of fluorescence in the measurement process. The fluorescence is isotropic, and thus *DT* and *TH* could be affected if a certain amount of fluorescence was induced.

5. Future Work

The 1st pilot study was finished in the end of 2016. Several reasons for non-equivalent *TH* in different measurement systems were investigated. There are currently many products labelled 'high haze' but no standard specifying the measurement method for this high range. The 2nd pilot study to investigate high haze samples has been approved by APMP with financial support, and will begin soon.

The research results of the first and second haze APMP TC initiative projects will be disseminated via a CIE reportership and potentially lead to a new CIE technical committee (TC). CMS/ITRI proposed the reportership (DR 2-79) of measurement of total transmittance, diffuse transmittance, and transmittance haze to CIE Division 2 in 2016, and it was passed with no objection. All participants will present the results of the pilot study to apply for a new TC in 2018.

6. References

[1] ASTM 2003 Standard test method for haze and luminous transmittance of transparent plastics American Society Testing and Materials ASTM D 1003

[2] ISO 1997 Plastics-Determination of haze of transparent materials International Organization for Standardization ISO 14782

[3] H.-L. Yu, C.-C. Hsiao and W.-C. Liu 2006 New apparatus for haze measurement for transparent media Meas. Sci. Technol. 17 N29-N36

[4] Hsueh-Ling Yu and Chin-Chai Hsaio 2009 Comparison of different measurement methods for transmittance haze Metrologia 46 S233-S237

[5] BS:2782 1992 Methods of Testing Plastics, Part 5: Method 521A: Determination of haze of film and sheet

[6] Goebel D G 1967 Generalized integrating-sphere theory Appl. Opt. 6 125-8

Polarization properties and microfacet-based modelling of white, grey and coloured matte diffuse reflection standards

T Quast[1], A Schirmacher[1], K-O Hauer[1] and A Koo[2]

[1]Physikalisch Technische Bundesanstalt, Braunschweig und Berlin, Germany
[2]Measurements Standards Laboratory, Lower Hutt, New Zealand

tatjana.quast@ptb.de

Abstract. To elucidate the influence of polarization in diffuse reflectometry, we performed a series of measurements in several bidirectional geometries and determined the Stokes parameters of the diffusely reflected radiation. Different types of matte reflection standards were used, including several common white standards and ceramic colour standards. The dependence of the polarization on the sample type, wavelength and geometry have been studied systematically, the main influence factors have been identified: The effect is largest at large angles of incidence or detection and at wavelengths where the magnitude of the reflectance is small. The results for the colour standards have been modelled using a microfacet-based reflection theory which is derived from the well-known model of Torrance and Sparrow. Although the theory is very simple and only has three free parameters, the agreement with the measured data is very good, all essential features of the data can be reproduced by the model.

1. Motivation

To obtain accurate and reliable results in diffuse-reflectometry experiments, one has to take all possible factors into account that have an impact on the experiment. One such factor on which we focus here is polarization. The polarization-related properties of the measurement apparatus as well as those of the sample under test have to be considered.

PTB's gonioreflectometer [1] has been equipped with a polarization-analyzer unit which allows to determine the Stokes parameters of the detected radiation [2]. We conducted a series of experiments to determine the main polarization dependence and influence factors and modelled the data.

2. Experimental details

The spectral radiance factor β of different reflection standards was determined using PTB's gonioreflectometer. A detailed description of the apparatus and the measurement procedure can be found in [1].

2.1. Matte diffuse-reflection standards

Various types of samples have been examined, among them some of the most common types of reflection standards. All samples had a diameter of 50 mm and a matte surface finish. Three different types of white reflection standards have been used: Opal glass, barium sulfate and ceramic. The spectral dependence of the polarization properties was examined using a set of coloured ceramic reflection

IOP Conf. Series: Journal of Physics: Conf. Series **972** (2018) 012024 doi:10.1088/1742-6596/972/1/012024

standards (light grey, dark grey, red, green and blue) within a comparison carried out in the project xDReflect [3]. Sandblasted aluminum was also investigated as a sample without sub-surface scattering contribution.

2.2. Stokes parameters and measurement procedure

A polarization-analyzer unit has been used to determine the Stokes parameters of the radiation, details of the setup and the measurement procedure can be found in [2]. The four Stokes parameters describe the polarization in terms of total intensity (I), linear polarization at 0°/90° orientation (M_0), linear polarization at +/-45° orientation (C_0) to the scattering plane, and circular polarization (S_0). The subscript "0" indicates that they are normalized by the intensity and converted to percent. The degree of polarization DOP is given by $\mathrm{DOP} = \sqrt{M_0^2 + C_0^2 + S_0^2}$. Since the light source used in the experiment is completely unpolarized, any detected polarization must originate from the sample under test.

For the measurements presented here, the geometry of illumination and detection has been varied (e.g. angle of incidence θ_i was kept constant at 0° while the angle of detection θ_r has been varied up to +/-75° in steps of 15° or vice versa. $\theta_{i/r} > 0$ means that the angle is measured from the surface normal (z-axis) in the direction of the positive x-axis, for $\theta_{i/r} < 0$ it is measured in the direction of the negative x-axis). The measurements were carried out at several fixed wavelengths in the visible spectral range.

3. Microfacet-based modelling

Data in this work has been described by the reflection model of Torrance and Sparrow [4]. It assumes that a rough surface is composed of mirror-like facets that have a distribution of slopes. There are two contributions to the reflected radiation. First, a specular contribution which is caused by specular reflection off those microfacets which have a suitable orientation. Polarization is included by using the Fresnel reflection coefficients to calculate the specular contribution. And second, a diffuse contribution which is assumed to be perfectly Lambertian.

The Torrance—Sparrow model shows a reasonable agreement with the measured data for most data sets, but it systematically fails if $\beta_p > \beta_s$ ($\beta_{s/p}$: spectral radiance factor for unpolarized incident radiation and detection sensitive only to s- or p-polarization, respectively). Thus, we decided to modify the Torrance—Sparrow model by coupling the specular and diffuse contributions. A similar approach has been used by Shirley [5], based on the idea that the diffuse contribution is modified to account for "specular losses": Any fraction of radiation, that has been reflected specularly, cannot contribute to the diffuse reflection. So, the diffuse contribution has to be decreased by the amount of specular reflection.

Since we found that for in-plane measurements only M_0 contributes to the state of polarization, it is sufficient to consider the radiance factor for s and p polarization, given by

$$\begin{aligned}\beta_{s/p} &= \beta_{s/p,\ \text{specular}} + \beta_{s/p,\ \text{diffuse}} \\ &= b\, R_{s/p}(n_1, n_2, \theta_i, \theta_r) \frac{G(\theta_r)}{\cos\theta_r} P(\theta_i, \theta_r) + a \cos\theta_i \left(1 - R_{s/p}(n_1, n_2, \theta_i, \theta_r)\right)\end{aligned} \tag{1}$$

where $R_{s/p}(n_1, n_2, \theta_i, \theta_r)$ are the Fresnel reflection coefficients for s and p polarized radiation. $G(\theta_r)$ is a shadowing function that accounts for masking and shadowing of the facets. In the original publication by Torrance and Sparrow, it is derived based purely on geometric optics. This results in a piecewise definition of G, which leads to discontinuities when fitting our data. Instead, we used a very simple, more empirical expression for $G(\theta_r)$, given by

$$G(\theta_r) = \frac{4}{\pi^2} \cdot \left(-\theta_r^2 + \frac{\pi^2}{4}\right) \tag{2}$$

$P(\theta_i, \theta_r)$ is the facet distribution which is assumed to be Gaussian

$$P(\theta_i, \theta_r) = \exp\left(-c^2 \cdot \left(\frac{\theta_i - \theta_r}{2}\right)^2\right) \tag{3}$$

IOP Conf. Series: Journal of Physics: Conf. Series **972** (2018) 012024 doi:10.1088/1742-6596/972/1/012024

The explicit wavelength dependence of the refractive indices and of the spectral radiance factor are omitted in this notation for better readability. To fit the model and to visualize the measured data in a less complex form, we used the ratio $\beta_{\text{ratio}} = \frac{\beta_s}{\beta_p}$ vs the angle $\theta_{i/r}$. The ratio makes it easier to identify small polarization effects ($\beta_{\text{ratio}} \approx 1$) compared to β_s and β_p. The refractive index of air was assumed to be unity ($n_1 = 1$). The refractive index of the ceramic reflection standards was determined by MSL, yielding values between 1.49 and 1.53. To model the data, we used a value of $n_2 = 1.5$. In this way, there are only three free parameters left for fitting: the relative amplitudes a and b, and the inverse width c of the facet distribution.

4. Results and discussion

4.1. Material dependence of polarization properties

Matte white standards are one of the most important classes of artefacts in diffuse reflectometry. They are used as reference artefacts in many applications. Thus, characterizing their polarization properties is of paramount importance. There are two mainly used bidirectional geometries in diffuse reflectometry of white standards. Either the radiation is incident perpendicular to the sample surface and detection takes place under 45° ($\theta_i = 0°, \theta_r = 45°$, denoted 0°:45°), or vice versa ($\theta_i = 45°, \theta_r = 0°$, denoted 45°:0°).

Table 1. Values for the Stokes parameter M_0 depending on the sample type and measurement geometry

Sample type	geometry	M_0 / %
Barium sulphate	0°:45°	-0.65
Barium sulphate	45°:0°	-1.23
Opal glass	0°:45°	2.7
Opal glass	45°:0°	-0.97
Ceramic	0°:45°	0.90
Ceramic	45°:0°	-1.48

All white standards that have been studied in this work exhibit a small, but non-zero amount of linear polarization that is introduced by reflection. The amount of linear polarization gets larger if either angle is increased. This effect is most pronounced for Opal glass. Even at 0°:45°, M_0 already has a value of about 3%, increasing up to 10% at $\theta_r = 75°$. For barium sulfate, M_0 is well below 4%, for ceramic even below 2%, even at large angles. The values obtained for both are summarized in Table 1. Measurements for Opal glass and ceramic were carried out at 700 nm, measurements for barium sulphate at 750 nm.

It was shown that for Opal glass at large detection angles, the reflected radiation is clearly partly polarized. The degree of polarization can become fairly large so that this effect cannot be neglected.

4.2. Spectral dependence of polarization properties

The wavelength dependence of the spectral radiance factor of the six ceramic standards (white, light grey, dark grey, red, blue and green) has been measured in a polarization-resolved way at MSL in the wavelength range between 370 nm and 720 nm. This allows to calculate the ratio β_{ratio} for the entire visible spectral range. The spectral measurements were carried out in both the geometries 0°:45° and 45°:0°. The results for all six standards are shown in Figure 1. The ratio β_{ratio} is plotted against the radiance factor for the unpolarized case, β_{unpol}, which is calculated as the average of the radiance factors for s and p polarization.

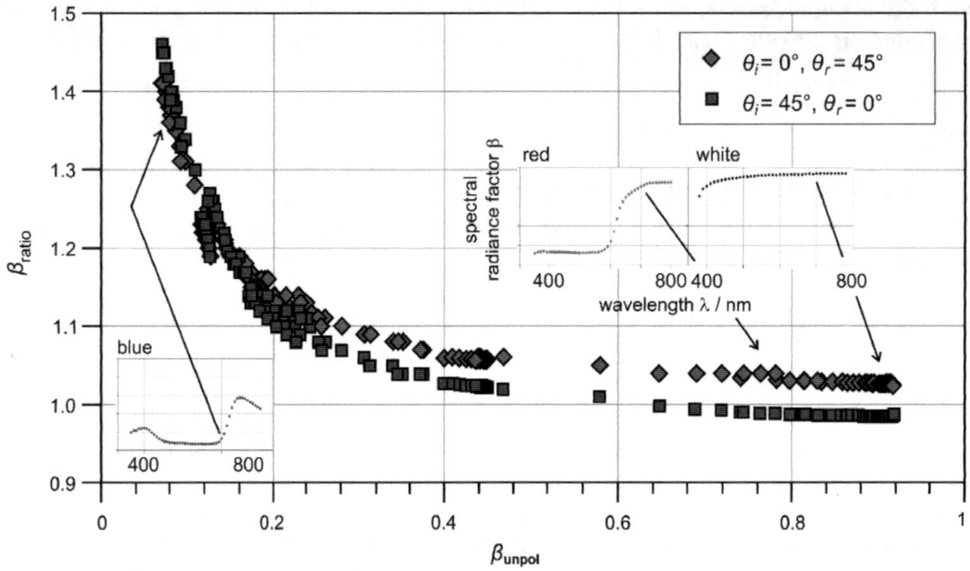

Figure 1. Spectral radiance factor ratio for all six ceramic reflection standards.

A strong correlation between the reflectance (given by the magnitude of β_{unpol}) and the polarization dependence can be seen: The lower the reflectance, the higher the ratio β_{ratio}, i.e. the larger the degree of polarization. This kind of observation has also been reported in other fields of reflectometry [6].

So, for this type of ceramic reflection standards, the reflectance indicates how large the influence of polarization is. For light samples and in the maxima of the reflection spectra, only a small influence can be expected. However, for dark samples and in the minima of the spectra, the influence can become very large, as ratios β_{ratio} up to 1.45 have been observed.

4.3. Geometry dependence of polarization properties

Since we have seen that for white standards, the measurements geometry plays a vital role for the polarization, we studied the influence of the geometry for the ceramic coloured standards as described in section 1.3. As for the white standards, we found that for larger angles, the influence of polarization increases. However, depending on the sample and the wavelength, the DOP gets considerably larger than for white standards. Values of up to 20% were found, especially for the dark samples. It was also observed that for most data sets, $\beta_{ratio} > 1$. For the white and red samples at large detection angles, we found $\beta_{ratio} < 1$. Some exemplary data sets are shown in Figure 2, together with the fit curves from the microfacet-based reflection model. It is apparent that there is a good qualitative agreement between the experimental data and the fit. All essential features of the data are well reproduced by the microfacet model. Although the agreement between the data and the model is remarkably good, considering that the model is very simple and only has three free fitting parameters, there are some shortcomings. They can be seen most clearly when we illustrate the dependence of the fit parameters on the magnitude of the DOP. The amplitude b of the specular term is almost constant for all data sets and does not depend on the DOP, so we refrain from depicting it graphically. The dependence of a (circles) and c (squares) is shown in Figure 3. Each symbol represents the fit parameter for a trace as given in Figure 2. The colour indicates the colour of the sample, the measurement wavelengths are stated explicitly next to the symbols.

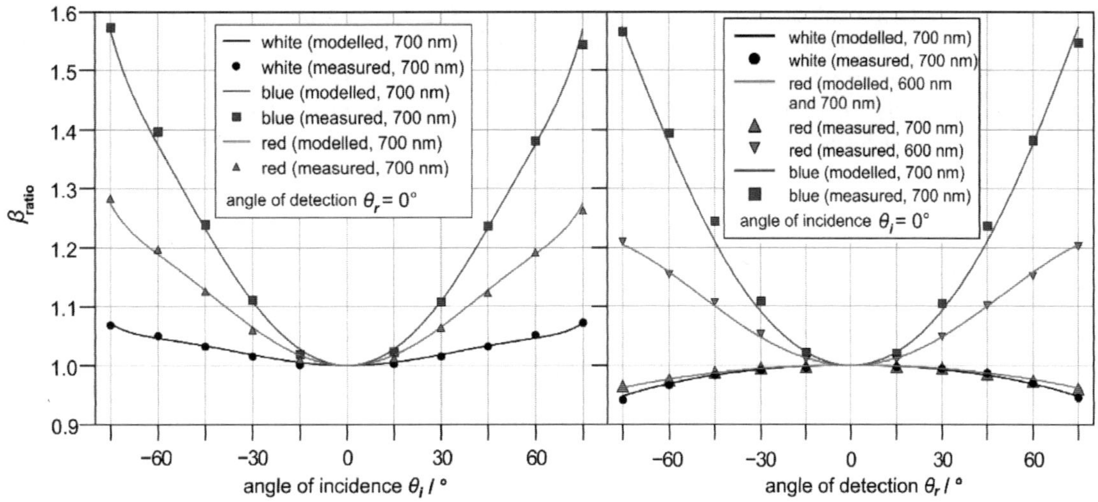

Figure 2. Spectral radiance factor ratio for different geometries. The experimental data are shown together with the best fit results of the microfacet-based reflection model.

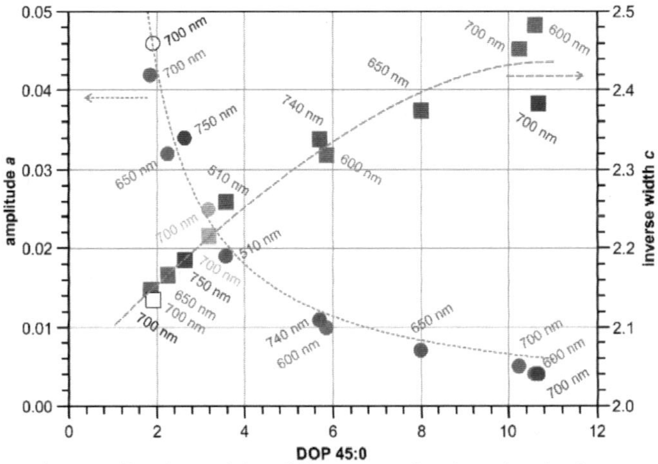

Figure 3. The best fit parameters for all data sets with variable angle of incidence, plotted against the DOP at $\theta_i = 45°, \theta_r = 0°$ for this data set. Circles represent the fit amplitude a (left axis), squares the inverse width c (right axis). The colour of the symbol indicates the colour of the sample, the measured wavelengths are given explicitly. The dashed and dotted lines are only shown to guide the eye.

The amplitude a of the diffuse contribution clearly decreases as the magnitude of the DOP increases. The higher the DOP, the smaller is the contribution of purely diffusely reflected radiation. It is, however, unexpected to see that the inverse width c of the facet distribution clearly correlates with the DOP. The larger the DOP, the larger c gets. This indicates the oversimplification of the model. At first glance, one would expect c to remain constant, since it is a parameter that describes the geometric characteristics of the reflecting surface. Even though different samples have been used, they are all comprised of the same ceramic and only differ in the dye used to set the color. Thus a similar surface characteristic for all samples can be assumed. It has been shown, though, that it can be useful to work with an "effective" surface parameter which depends on the measurement geometry. He et al. [7] used an effective surface roughness in their model since the roughness appears to decrease for oblique angles due to shadowing of the facets. The wavelength also appears to have an influence on c. The scattering behavior depends on the size of the surface structure relative to the wavelength. This is also included in He's model [7]. This approach has not been pursued in the model presented here, which could at least partly explain the variation of c.

Another reason might be that our model is based purely on geometrical optics. Scattering of radiation from a rough surface is a far more complex process, effects like interference on small-scale roughness or subsurface scattering can play an important role. None of these are included in the model used here. We can expect subsurface scattering to have a large impact on the reflection properties of ceramic standards. When measuring a matte sample made of sandblasted aluminum in the same geometries as the ceramic standards, almost no effect of polarization was found ($M_0 < 1\%$ even at large angles). For aluminium, subsurface scattering can be ruled out since the radiation does not penetrate the bulk of the material. In turn, this indicates that the large degree of polarization found for the ceramic standards can be connected to subsurface scattering. It probably originates from scattering at the interface between the coloured glaze layer and the ceramic substrate.

5. Conclusion

In this work, we studied the polarization properties of diffusely reflected radiation of matte diffuse reflection standards, including different types of white and coloured standards. For the ceramic coloured standards, we analysed the wavelength and geometry dependence of the polarization properties. We modelled the data using a microfacet-based reflection model that was derived from the model introduced by Torrance and Sparrow.

Only in-plane geometries have been considered, and we found that the only Stokes parameter for linear polarisation M_0 has a measurable value. All white standards we studied showed a small, but nonzero DOP. The effect was most pronounced for Opal glass at large detection angles, yielding values of up to 10%. For barium sulfate and ceramic, the values are smaller, but in general not negligible. When studying the wavelength dependence of the polarization of the ceramic colour standards, we found a clear correlation between the magnitude of the reflectance and the DOP. The lower the reflectance, the higher the DOP, thus the larger is the influence of polarization. At selected wavelengths, we studied the geometry dependence. The larger the angle which is varied (either incidence or detection), the higher the DOP. In short, for dark samples at oblique angles, the DOP can easily take values of up to 20%. For light samples at almost perpendicular incidence or detection angle, the impact of polarization is much smaller, but still nonzero.

We modelled the data, using a modified Torrance—Sparrow reflection model, in which the diffuse contribution has been modified to account for "specular losses". Despite being very simple and only having three free fitting parameters, the model yields a remarkably good agreement with the measured data. However, some shortcomings are apparent since some of the variation in the best fit parameters cannot be explained satisfactorily.

Even for such basic artefacts as matte diffuse reflection standards, the influence of polarization has to be considered carefully, as it depends on a number of parameters: the sample material, the wavelength, the measurement geometry, the polarization properties of the measurement setup and the targeted uncertainty.

References

[1] Hünerhoff D, Grusemann U and Höpe A 2006 *Metrologia* **43** 11-16
[2] Quast T, Hauer K and Schirmacher A 2016 *Proc 4th CIE Expert Symposium on Colour and Visual Appearance* 30-39
[3] Strothkämper C, Ferrero A, Koo A, Jaanson P, Ged G, Obein G, Källberg S, Audenaert J, Leloup F, Martinez-Verdu F, Perales E, Schirmacher A and Campos J 2017 *Applied Optics* 1996-2006
[4] Torrance K and Sparrow E 1967 *JOSA* **57** 1105-14
[5] Shirley P, Hu H, Smits B and Lafortune E 1997 *Proc of the fifth Pacific Conference on Computer Graphics and Applications* 40-49
[6] Umov N 1905 *Physik Z* **6** 674-676
[7] He X, Torrance K, Sillion F and Greenberg D 1991 *Computer Graphics* **25** 175-186

Investigation of converging and collimated beam instrument geometry on specular gloss measurements

Joanne C. Zwinkels[1], Éric Côté[1] and John Morgan[2]

[1]National Research Council of Canada, Ottawa, Canada
[2]University of Waterloo, Waterloo, Canada

Joanne.zwinkels@nrc-cnrc.gc.ca

Abstract. Specular gloss is an important appearance property of a wide variety of manufactured goods. Depending upon the application, e.g. paints, paper, ceramics, etc. different instrument designs and measurement geometries are specified in standard test methods. For a given specular angle, these instrument designs can be broadly classified as converging beam (TAPPI method) and collimated beam (DIN method). In recent comparisons of specular gloss measurements using different glossmeters, very large standard deviations have been reported, well exceeding the manufacturers claims. In this paper, we investigate the effect of instrument beam geometry on gloss measurements. These results indicate that this difference in beam geometry can give the magnitude of gloss differences reported in these comparisons and highlights the importance of educating the user community of best measurement practices and obtaining appropriate traceability for their glossmeters.

1. Introduction

In a recent investigation of the inter-instrument agreement of specular glossmeters conducted as part of the European Metrology Research Programme (EMRP) project entitled "Multidimensional Reflectometry for Industry" (xDReflect), it was reported that the reproducibility between 6 different commercial glossmeters on 25 different painted gloss samples exceeded the threshold gloss unit (GU) values given in ASTM D523-14 for more than half of the comparison samples [1]. These threshold values are already extremely high ranging from 3.5, 6.4 and 7.2 GU for measurement geometries of 60°, 20° and 85°, respectively.

At the National Research Council of Canada (NRC), we have developed reference instruments to provide traceable gloss measurements in accordance with different standard test methods. This includes the well-established NRC Reference Glossmeter for specular gloss measurements at 20°, 60° and 85° geometries in accordance with ISO 2813 and ASTM D523 [2,3] which has been providing high-accuracy specular gloss calibration services for more than 30 years and the more recently developed NRC Reference Goniospectrophotometer (GSP) [4,5] which is a highly versatile instrument that can be readily re-configured to provide traceable specular gloss measurements in accordance with several standard test methods for different applications that specify either a collimated beam (so-called DIN method) or converging beam (so-called TAPPI method) geometry. To the best of our knowledge, this is the only reference glossmeter that provides traceable calibrations in a converging beam geometry.

IOP Conf. Series: Journal of Physics: Conf. Series **972** (2018) 012025 doi:10.1088/1742-6596/972/1/012025

Even for a given application, such as the paper industry, there are different standard test methods for specular gloss measurements at a given specular angle that specify different instrument beam geometries. For coated and supercalendared papers, their specular gloss is typically measured at 75° incident angle using: ISO 8254-1:75° gloss with a converging beam (TAPPI method) or ISO 8254-2: 75° gloss with a parallel beam (DIN method). For other types of glossy papers, such as highly varnished or waxed papers, they are typically measured at 20° incident angle using ISO 8254-3: 20° gloss with a converging beam: TAPPI method or TAPPI T653. The most widely used gloss standard test methods are ASTM D523 and ISO 2813 which cover not only the measurement of specular gloss of paint samples but general non-metallic materials. Since these standards include measurements for a 20° specular angle and allow for either a collimated or converging beam geometry, they could also be used for measurements of certain types of glossy paper samples.

The choice of whether a collimated or converging beam geometry is most appropriate depends upon the nature of the sample. For uniform glossy samples, a collimated beam geometry is generally recommended since it provides better uniformity of the reflected rays received in the detector aperture. On the other hand, for non-uniform or textured glossy samples, where it is desirable to obtain information about the spatial or orientational dependence of the sample gloss, a converging beam geometry is recommended since it provides better uniformity in extent of the angular divergence that is accepted in the measurement. It is also important to note that a glossmeter that conforms to a converging beam geometry must meet additional specifications in terms of spatial uniformity for both the source and receptor aperture, that are not required for a glossmeter conforming to a collimated beam geometry. For this reason, most commercial glossmeters that conform to ASTM D523 or ISO 2813 have a collimated beam geometry.

To the best of our knowledge, there is no published information on the impact on specular gloss measurements for glossy samples measured at a given specular angle for different instrument beam geometries (converging and collimated). For this reason, we began a study with the NRC GSP to study this effect for glossy paper samples at the aforementioned 75° and 20 °geometries.

2. Comparison Samples

Glossy paper samples were obtained from the Collaborative Testing Services (CTS) Paper Program interlaboratory testing of specular gloss measurements. For the measurements at 75° geometry, we used a medium (GU53) and high gloss (GT53) white paper sample, comprising 10 specimens each, obtained from CTS in ~2002, when the GSP was being validated for 75° gloss in accordance with TAPPI T480. For the measurements at 20° geometry, we obtained a medium gloss sample (RG01) comprising 30 specimens.

For each of the specimens, one measurement was to be performed with the centreline of the light beam in the plane perpendicular to the sheet and parallel to its long direction. The sheet was then rotated in its plane 180° and measured again. The mean results for these 2 measurements were reported, giving a total of either 10 or 30 results, depending on test sample.

Coincidentally, CTS recently conducted a 20° gloss comparison using the sample RG01, on behalf of TAPPI, to determine if the technical parameters with regard to acceptance angles could be loosened for the TAPPI method T653: *specular gloss of paper and paperboard at 20°*. CTS kindly provided us these comparison results which are plotted below in a histogram to show the distribution of the 30 data points for the 14 participating labs. There is a bimodal distribution with one cluster of labs reporting a value ~30-40 GU and the second cluster of labs reporting a value of ~50-60 GU, with a grand mean of 44.9 GU and standard deviation (S.D.) of 8.7 GU.

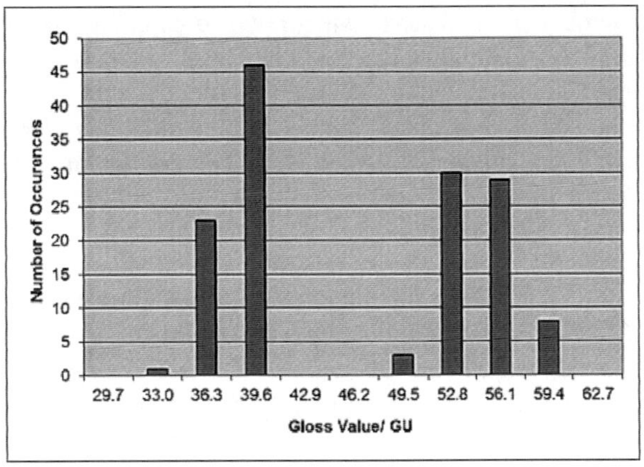

Figure 1. CTS 20° TAPPI gloss inter-lab results reported for glossy paper sample, RG01 (14 participating labs).

3. NRC Gloss Measurement Results

The NRC GSP and these 3 glossy white paper samples were used to investigate the impact of going from a collimated to a converging beam geometry for both 20° and 75° specular angles. To test the precision of these gloss measurements, two independent runs were carried out, reconfiguring and realigning the instrument between repeat runs. Figure 2 shows a schematic drawing of the GSP with an exploded view of the key components that need to be changed when converting the instrument to a different beam geometry; these are the source field stop, receptor aperture and an optional source objective lens assembly (not shown here) to convert from collimated beam (shown here) to converging beam. To validate the instrument performance, two NRC high gloss quality system check standards were used and measured at the beginning, middle and end of each sample run to check for any temporal drift in the gloss measurements. These quality system check standards typically gave agreement within 0.2% with their historical results, indicating a very high level of instrument repeatability.

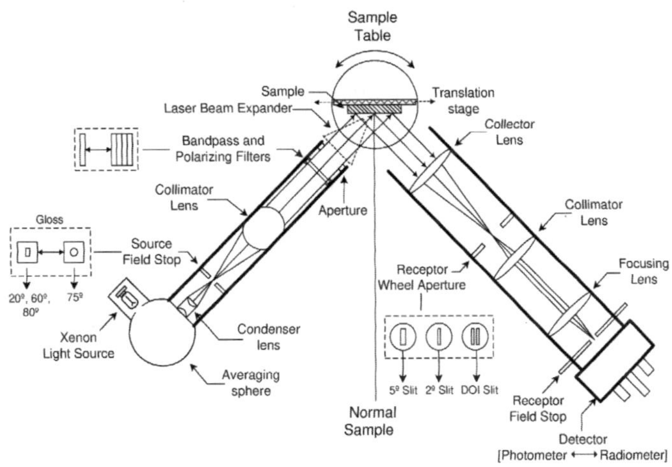

Figure 2. Schematic drawing of NRC Reference Goniospectrophotometer (GSP) indicating the key components (in dashed rectangles) that are changed when converting the instrument from collimated (shown here) to converging beam geometry. An optional lens assembly is also inserted in the source path between the source field stop and the sample stage.

The NRC primary gloss standard is a high optical quality quartz prism whose gloss values are calculated from its known optical constants and the Fresnel relations for the given specular angle and referenced to the refractive index value specified in the relevant gloss standard test method [3]. It should be noted that the primary gloss standard is defined with a different refractive index, n for these two methods; n=1.540 (TAPPI) and n=1.567 (DIN). The calculated values for the NRC quartz wedge standard are included here in the NRC tables of gloss results. These results are given in Table 1 for the 75° geometry for (a) TAPPI and (b) DIN methods, respectively, and in Table 2 for the 20° geometry. For the 20° geometry, the NRC gloss measurement results are also compared with the grand mean and standard deviation (S.D.) results from the CTS 20° TAPPI round-robin comparison.

It can be seen that for the 75° geometry, the gloss differences are a factor of 3 to 6 times the measurement precision, whereas for the 20° geometry, the gloss differences are more than a factor of 20 times. In comparing the NRC gloss results in Table 1 and Table 2, it is interesting to note when going from the converging to the collimated beam instrument geometry, in the case of 75°, the NRC gloss results for both the medium and high gloss paper sample increased, whereas for 20°, the NRC results for the medium gloss paper sample decreased. This suggests that it is not possible to necessarily predict and correct the measured gloss values for the impact of change in beam geometry and that this impact would need to be evaluated for each different gloss geometry and type of specimen.

Table 1. Comparison of measured specular gloss, expressed in GU values, at 75° on NRC GSP for a medium and a high gloss paper sample for: (a) a converging beam (TAPPI method) and (b) a collimated beam (DIN method) geometry.

ID	Run 1 (GU)	Run 2 (GU)	ID	Run 1 (GU)	Run 2 (GU)
GU53	33.03	31.03	**GU53**	37.94	38.73
S.D.	±0.75	±0.73	S.D.	±0.82	±0.83
GT53	66.43	65.30	**GT53**	73.38	73.98
S.D.	±2.00	±2.36	S.D.	±1.94	±1.91
Quartz Standard	92.65	92.65	**Quartz Standard**	94.22	94.22

(a) TAPPI method (b) DIN method

Table 2. Comparison of measured specular gloss, expressed in GU values, at 20° for paper sample RG01 on NRC GSP for: (a) a converging beam (TAPPI method) and (b) a collimated (DIN method) geometry and compared with the results reported by CTS for 20° gloss (TAPPI) comparison for 14 labs.

Gloss Standard: (GU value)	**Run #**	**NRC results: RG01**		**CTS results (14 labs): RG01**	
		Gloss (GU)	**S.D. (GU)**	**Gloss (GU)**	**S.D. (GU)**
TAPPI	1	36.02	0.65	44.90	8.70
Quartz Standard: (77.39)	2	34.51	1.22		
DIN	1	58.98	0.61		
Quartz Standard: (71.72)	2	58.96	0.76		

IOP Conf. Series: Journal of Physics: Conf. Series **972** (2018) 012025 doi:10.1088/1742-6596/972/1/012025

4. Conclusions

The results of this gloss comparison using the NRC reference goniospectrophotometer configured in both a converging and collimated beam geometry for measuring specular gloss of paper samples at 75° and 20° geometries showed that for the paper samples studied here, for a given specular angle and beam geometry, the measurement reproducibility was very good, ranging from 0.6 -2.4 GU. However, for both 75° and 20° geometries, for a change in the beam condition, the measurement reproducibility significantly deteriorated. In the case of both the medium (GU53) and high gloss (GT53) coated paper samples measured at 75° geometry, the differences in the measured gloss values for the two different methods (6.3 –7.8 GU) were a factor of ~4 to 8 times larger than the measurement reproducibility for a given beam condition (0.7 -2.4 GU). In the case of the medium gloss paper sample (RG01) measured at 20° geometry, the NRC results showed a very large gloss difference of 23.4 GU in going from the converging (TAPPI) to the collimated (DIN) beam method compared with an excellent measurement reproducibility of 0.6-1.2 GU for a given beam condition . It is interesting to note that the NRC mean result for sample RG01 using these two methods (47.1 GU) is in very good agreement with the grand mean result of 44.9 GU that was reported for the CTS 20° TAPPI round-robin study using this glossy paper sample (see Figure 1).

These results strongly suggest that differences in beam geometry can account for the very large inter-instrument differences reported in recent gloss comparison studies that greatly exceed both gloss manufacturer and test method specifications for repeatability and reproducibility. These NRC results also highlight the need for better education of the user community on the traceability of their commercial glossmeters and to confirm that it strictly conforms to the geometric, spectral and photometric conditions of the specified gloss standard test method, including the conditions of the incident beam (converging or collimated).

References

[1] F..B. Leloup, J. Audenaert, and P. Hanselaer, Investigation of the inter-instrument agreement of specular glossmeters, in Proceedings of 28th CIE session, CIE 216:2015, 355-362, 2015.

[2] W. Budde, A reference instrument for 20°, 60°, and 85° gloss measurements, Metrologia, 16, 1-5, 1980.

[3] J.C. Zwinkels and M. Noël, Specular gloss measurement services at the National Research Council of Canada, Surface Coatings International, 78, 512-515,1995.

[4] J. Liu, M. Noël and J. Zwinkels, Design and characterization of a versatile reference instrument for rapid, reproducible specular gloss measurements, Appl. Opt., 44, 4631-4638, 2005.

[5] M. Noël, J. Zwinkels and J. Liu, Optical characterization of a reference instrument for gloss measurements in both a collimated and a converging beam geometry, Appl. Opt., 45, 3712-2720, 2006.

IOP Conf. Series: Journal of Physics: Conf. Series **972** (2018) 012026 doi:10.1088/1742-6596/972/1/012026

Development of a new linearly variable edge filter (LVEF)-based compact slit-less mini-spectrometer

Khaled Mahmoud[1,2,3,*], Seongchong Park[1], and Dong-Hoon Lee[1]

[1]Division of Physical Metrology, Korea Research Institute of Standards and Science (KRISS), Daejeon, 34113, Rep. Korea.

[2]National Institute for Standards (NIS), PO Box 136 Giza, Code No 12211, Egypt.

[3]National Measurement and Calibration Center (NMCC), SASO, PO Box 3437 Riyadh 11471 Saudi Arabia.

[*]k.abdelftah@saso.gov.sa/khaled55eg@yahoo.com

Abstract. This paper presents the development of a compact charge-coupled detector (CCD) spectrometer. We describe the design, concept and characterization of VNIR linear variable edge filter (LVEF)- based mini-spectrometer. The new instrument has been realized for operation in the 300 nm to 850 nm wavelength range. The instrument consists of a linear variable edge filter in front of CCD array. Low-size, light-weight and low-cost could be achieved using the linearly variable filters with no need to use any moving parts for wavelength selection as in the case of commercial spectrometers available in the market. This overview discusses the main components characteristics, the main concept with the main advantages and limitations reported. Experimental characteristics of the LVEFs are described. The mathematical approach to get the position-dependent slit function of the presented prototype spectrometer and its numerical de-convolution solution for a spectrum reconstruction is described. The performance of our prototype instrument is demonstrated by measuring the spectrum of a reference light source.

1. Introduction

Optical spectrometers are widely used in variety of applications in many fields relating to analysis of light sources and luminaries as well as analyzing distant objects in space missions, earth observation applications among others. Where today is the age of scaling-down and miniaturization, so the need of mini-spectrometer arises and becomes of great interest. Moreover, the need of miniaturization appears and becomes essential for some industrial and scientific applications. For instance, in space missions' applications, instruments need to be compact and small in weight. Hence, developing a compact and light spectrometer is of great interest for scientists in variety of fields [1,2]. Many approaches tried to address this need by constructing mini-spectrometers in which, for example, the traditional optical elements, e.g. gratings and prisms, were replaced by optical filters. The main drawback of a traditional filter instrument is that it is always dedicated to a specific task and, as far as it is concerned, no other spectral data exist. Recently, new technology provides a filter known as a linear variable optical filter (LVOF), which is capable of providing a set of continuously changing wavelengths across its length.

A very compact and simple micro-spectrometer can be produced by combining one of these filters with a CCD/CMOS array. There are different types of filters that can be used in spectrometers, e.g. bandpass transmission filters [3-6].

With this type of mini-spectrometers, the peak/edge wavelength of the transmission filter is moving along one direction of the filter, such that each line of a two-dimensional array detector, equipped with this filter, will detect radiation in a different wavelength band. One of the candidates from this type is the bandpass filter based spectrometer. Using graded bandpass filters has some drawbacks, e.g. limited resolution, so using a linear variable edge filter instead with a well-prepared mathematical algorithm can have significant advantages due to our concept.

In this paper, we describe our new instrument and its basic concept, we characterize our instrument to test the behaviors which are directly related to our application, e.g. measuring transmittance at different positions of the filter, test edge steepness, test the linear relationship between edge wavelength and position throughout the filter and finding the slit functions at different positions throughout the active area of our instrument. At end, we show preliminary results of our initial measurements, by applying our model with de-convolution algorithm, of reference light source with comparing the results with those of reference traditional spectrometer.

2. Working concept

2.1. Construction of prototype

Our approach to develop the prototype LVEF- based spectrometer is based on two main parts, first the design and construction of the device using a LVEF filter in front of CCD sensor and second to develop the optimal algorithm for treatment of the output digital image data. The off-the-shelf linear variable filters can cover the spectral region ranging from 350 nm to 1000 nm [7,8]. This approach permits wide operating wavelength ranges but, wider is the spectrum; greater are the difficulties of instrumentation design, fabrication and mathematical algorithm.

As shown in figure 2, our hardware module comprises the linearly variable edge filter and CCD monochrome (B/W) digital sensor (Sony, model ICX205AL, 1280 × 1024 pixels, pixel size 4.65 μm × 4.65 μm with 8 bit analog-to-digital converter). On top of the CCD array, we place LVEFs with some main characteristics, e.g. having "very steep edges", "~90% transmission or more", "low ripple", better than OD3 blocking (short wave rejection). We cut the main filter, which covers the spectral range from 300 nm to 850 nm, into 4 identical pieces to construct four replaceable filters; A: 300-430 nm, B:440-570 nm, C:580-710 nm and D:720-850 nm set on the sensor fitting its transversal dimension (see Figure 1). To achieve the best measurements, the variable filter should be mounted as close as possible to the sensitive area of the sensor, this helps to have the optimum spectral resolution and avoid any light divergence. To reduce the effects of the back reflection, an antireflection coating deposited on the sensor sensitive area is necessary. We made a special designed filter holder so that it can accommodate one filter in a time to fulfil the important criteria as mentioned above, e.g. installing each filter securely and easily as close as possible to the sensor surface and to make it easy for each filter to be changed in order to scan through its specified wavelength range. We described the main idea of this work and its flow chart in [9]. The main concept and the mathematical model of our approach will be explained in the next section.

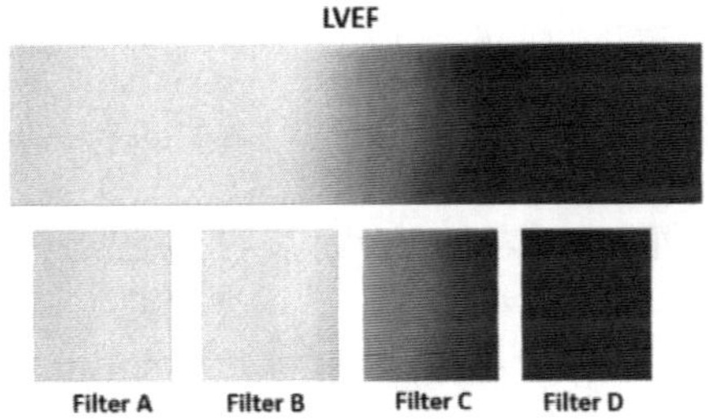

Figure 1. Main LVEF and its components result from fine cutting process.

IOP Conf. Series: Journal of Physics: Conf. Series **972** (2018) 012026 doi:10.1088/1742-6596/972/1/012026

2.2. Concept of the mathematical algorithm

In literature many algorithms for data treatment and image post-processing are described. The main issue with the available algorithms is that it always addresses specific cases. Using optimum mathematical software algorithms, it is possible to obtain correct spectral information about the samples under study. Generally, spectral data processing can be divided into two successive steps: pre-processing of raw data and extracting the necessary information. Superimposed or convolved spectral components can be only partly separated by spectral data processing usually referred to as de-convolution. De-convolution is the transformation of the measured spectrum into the one which is as close as possible to the true spectrum [10-13].

Figure 2 shows the schematic diagram of the hardware module (prototype) which is supposed to receive the radiation from the test source. It is shown that at different points throughout the horizontal axis the different pixels will receive light corresponding to different "cut-on" band and the adjacent pixels receive another band and so on. Let's suppose that, each channel (bundle of detectors/pixels) in the array will receive some signal yi corresponding to its position in the horizontal axis within the array, this defines which spectral component this channel will receive, so the mathematical concept for the model can be understood due to the following steps.

The output signal y_i of each sensor D_i (i = 1 ... n) can be expressed as

$$y_i = \int S(\lambda) \cdot R(\lambda) \cdot T_i(\lambda) d\lambda \quad (1)$$

where S(λ) denotes spectral power of the input radiation, R(λ) the spectral responsivity of the sensor (which is assumed to be same for all the sensors in the array after being corrected, with the normal procedures, for the photoresponse non-uniformity), and $T_i(\lambda)$ the spectral transmittance of the linearly variable edge filter (LVEF) at the position of the sensor D_i. Without loss of generality, we assume that the LVEF is a long-pass filter attached to the sensor array with the shortest cut-on wavelength at D_1 and the longest cut-on wavelength at D_n.

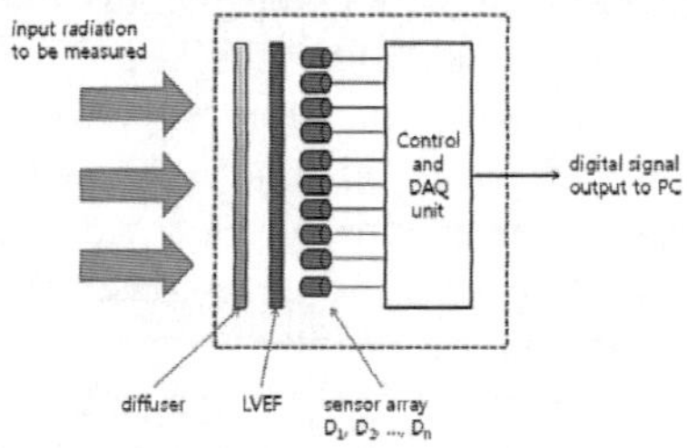

Figure 2. Schematic diagram of the hardware module.

We consider the difference δ_i (i = 1 ... n-1) between two sensors D_i and D_{i+1} as the raw data of the spectrometer:

$$\delta_i \equiv y_{i+1} - y_i = \int S(\lambda) \cdot R(\lambda) \cdot \left[T_{i+1}(\lambda) - T_i(\lambda)\right] d\lambda \quad (2)$$

Figure 3 shows the change of positions of D_i throughout the captured frame to make the idea visible and showing the direction of change of transmission change throughout the filter/sensor combination.

For the LVEF, each position on the filter can be assigned to a cut-on wavelength of the long-pass filter edge with a linear function. We can define the "slit function" $\Lambda(\lambda,\lambda_i)$ of this spectrometer as

$$\Lambda(\lambda,\lambda_i) \equiv R(\lambda)\left[T_{i+1}(\lambda) - T_i(\lambda)\right] \quad (3)$$

with λ_i is the cut-off wavelength assigned to the filter position at the sensor D_i. Then, Eq. (2) can be re-written as

$$\delta(\lambda_i) \equiv \delta_i = \int S(\lambda) \cdot \Lambda(\lambda,\lambda_i) \cdot d\lambda \quad (4)$$

Note that the slit function $\Lambda(\lambda,\lambda_i)$, which is the characteristic property of the spectrometer instrument, can be experimentally determined at the position of each sensor D_i (i = 1 ... n-1) by using a tunable monochromatic source. We will show the experimental results of studying the slit function and its position dependency in subsequent section.

Note also that the slit function can be, in general, different for each sensor position so that $\Lambda(\lambda,\lambda_i)$ forms a two-dimensional matrix. The task of the Data Acquisition (DAQ) unit is to calculate and plot $S(\lambda)$ from the measured raw data array $\delta(\lambda_i)$ by using the slit function matrix $\Lambda(\lambda,\lambda_i)$ of the instrument via a numerical de-convolution.

Now we consider an idealized special case where the spectral transmittance $T_i(\lambda)$ satisfies the following relation:

$$T_{i+1}(\lambda) = T_i(\lambda - \delta\lambda) \tag{5}$$

This means that the spectral transmittance of the LVEF shifts by a constant amount on the axis of wavelength without any change of its shape, as the sensor position varies by one step. Moreover, the responsivities of the sensors are wavelength-independent. Then, the shape of the slit function in Eq. (3) will be a fixed symmetric function with the symmetry center at the cut-on wavelength λ_i, which moves by a fixed value of $\delta\lambda$ as the sensor position changes by one step. We can simplify Eq.(4) for this case of the constant slit function to

$$\delta(\lambda_i) = \int S(\lambda)\cdot\Lambda(\lambda - \lambda_i)\cdot d\lambda \tag{6}$$

Eq. (6) has the same form as a scanning grating spectrometer with a position-independent slit function, and its de-convolution solution is well described in the literature.

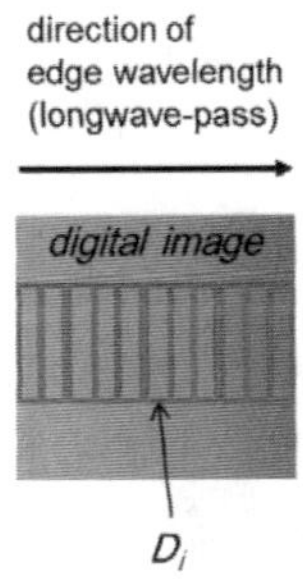

Figure 3. Spectral frame showing transmission channels throughout the filter/sensor combination.

3. Experimental results

3.1. LVEF characterization

3.1.1. Spatial transmittance measurements

Figure 4 shows schematically the setup for transmittance measurement of our filters at different positions throughout the filter surface, noted in eq 1 as T_i. Following our concept to determine the slit function at different positions, we should first determine the value of T_i and ΔT_i in order to determine the "slit function" $\Lambda(\lambda,\lambda_i)$ at different positions "i" considering that the responsivity is uniform throughout the whole sensor active area.

The apparatus to measure the graded transmission comprises QTH based monochromatic light source and CCD sensor. Light source part constitutes QTH lamp, double-grating monochromator; for minimizing spectral stray light and optimizing the resolution with keeping spectral bandwidth within appropriate limits and achieving excellent spectral purity, and set of order sorting filters for out-of-band stray rejection. Six-inch integrating sphere painted with PTFE-based high reflective material used with one-inch output and entrance ports and half-inch north-pole port for installing reference

IOP Conf. Series: Journal of Physics: Conf. Series **972** (2018) 012026 doi:10.1088/1742-6596/972/1/012026

detector. Collimating and focusing fore-optics comprising two off-axis focusing mirrors and set of stops used for imaging the exit slit at integrating sphere entrance plane and for beam shaping. Integrating sphere, working as perfect diffuser, provides uniform illumination at filter/sensor plane to keep the general principle of uniform response of the CCD sensor valid.

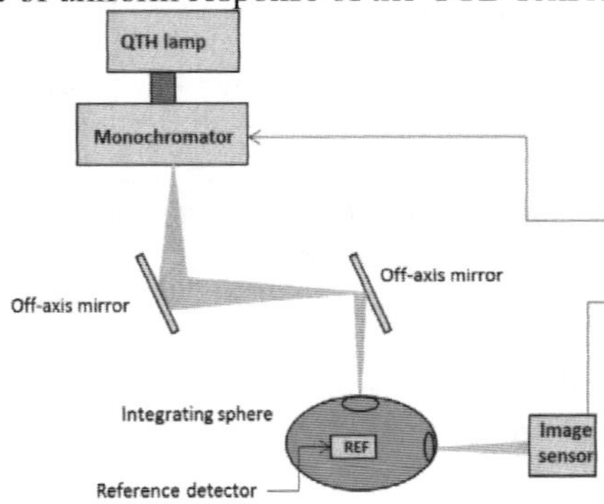

Figure 4. Setup for spectral transmittance measurements.

Figure 5 shows some measurements of transmittance at different horizontal positions which has been recorded by the CCD sensor at different channels (regions of interest –ROIs-). These measurements made by using a tunable monochromatic source described above. The steepness of transmitted spectra at different spatial positions are of 10 to 20 nm, which are reasonable to set our concept into practice with a good accuracy. As discussed earlier, we cut the filters into small sizes to have them nearly fitting the sensor area. Each filter covers smaller spectral range with the set of replaceable four filters allowing our device to cover a wide spectral range from 300 nm to 850 nm. We present transmittance data of three of our filters in figure 5 showing the movement of the edges throughout each filter.

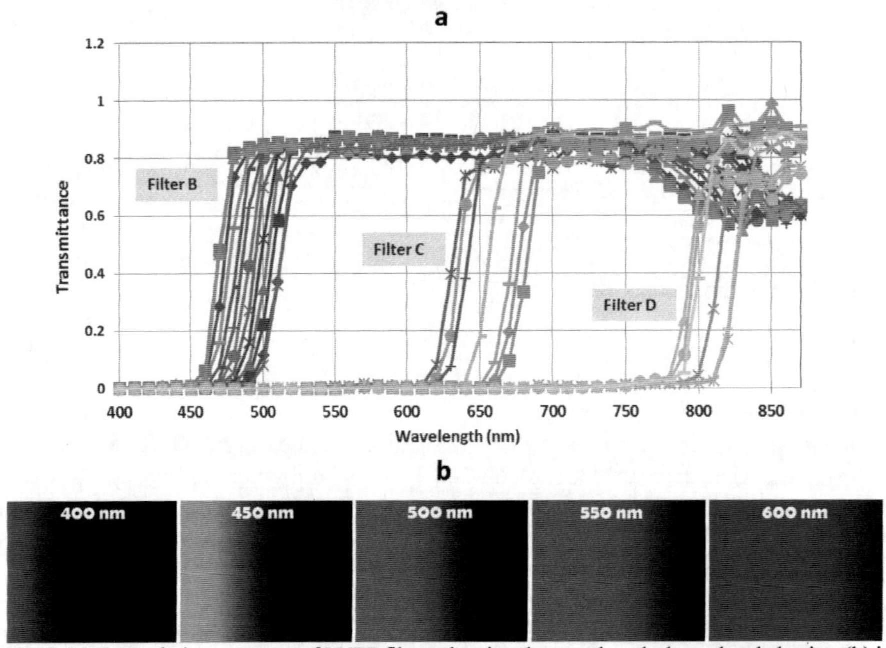

Figure 5. (a) Transmission spectrum of LVEF filters showing the wavelength-dependent behavior, (b) image of the filter on the CCD illuminated at different wavelengths.

In Figure 6 the variation of the edge wavelength along the surface of three of the used filters over 3 mm in each case is shown. Linear relationship between position and wavelength has been proved with these measurements. Knowing that the linear movement of the edges throughout the filter is important factor to be proved, where in our concept in developing the numerical algorithm for collected images' post-processing, the relationship between the spatial position at the filter and the wavelength should be linear. To present this relationship, we plot the sum of pixel counts (intensities) in adjacent ROIs with the same sampling and steps of 100 pixels (~ 0.5 mm) in the transversal direction as a function of position throughout the filter. The linear relationship in this case proves the linear displacement of edge wavelength throughout the filter, which is the main advantage of linearly variable filters technology.

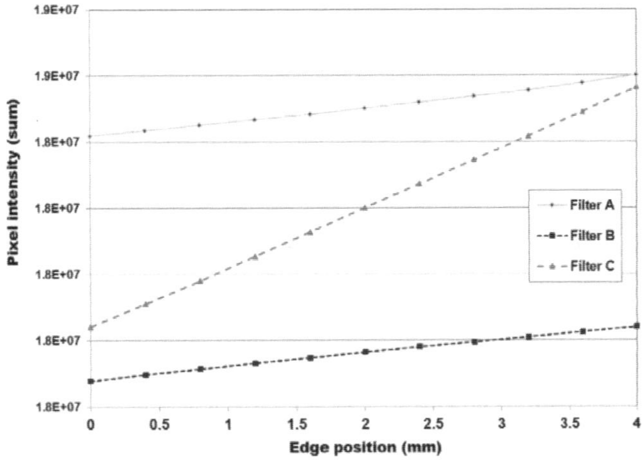

Figure 6. Linearity of the edge displacement along the direction of variation presented for filters A, B and C.

3.1.2. Determination of the slit functions

To determine slit function of our instrument we should have the spectral transmittance T(λ) and responsivity R(λ) while knowing T(λ) at each ROI with index "i" (see figure 3). Then, according to the mathematical model described in the previous section we can calculate the slit function at each ROI which denoted as $\Lambda(\lambda,\lambda_i)$ in eq 3.

Figure 7 represents set of calculated slit functions with different sampling intervals throughout different filters. Change the sampling by changing the position of "i+1" in eq 3 with respect to "i" yields different bandwidth (FWHM) at this position "i".

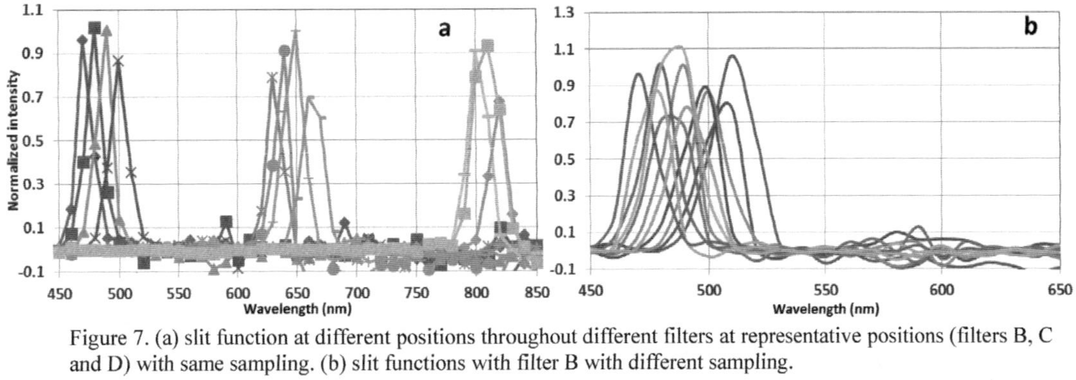

Figure 7. (a) slit function at different positions throughout different filters at representative positions (filters B, C and D) with same sampling. (b) slit functions with filter B with different sampling.

In figure 8, we show the shape and spectral width of the slit function for one position at two of the filters, while having different sampling in each case with respect to the other. It should be noticed that, the shape of the slit function is ideal; triangular profile. It is clearly shown that by changing sampling,

IOP Conf. Series: Journal of Physics: Conf. Series **972** (2018) 012026 doi:10.1088/1742-6596/972/1/012026

we can different "FWHM" of our slit function. This clearly shows that, in our instrument, we can have different flexible and tunable wavelength band through mathematical modeling. That's one of the big advantages of our mini-spectrometer. With mathematical modeling, we can change sampling to get boarder band for getting higher signal-to-noise ratio (SNR). The numerical algorithm is a key factor in the success of any model in all similar cases [10-13].

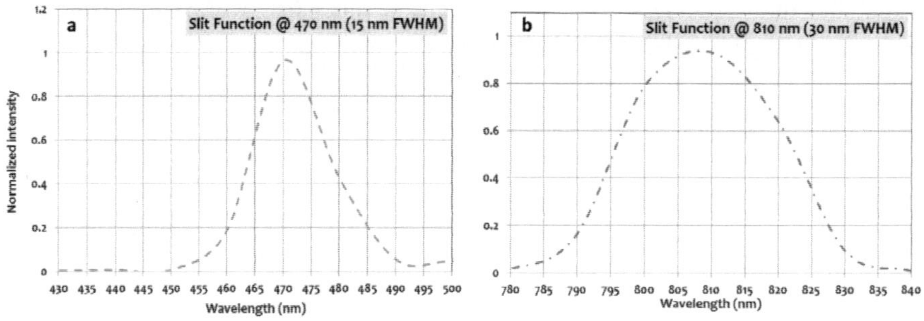

Figure 8. (a) Slit function at 470 nm with FWHM of 15 nm, filters B. (b) slit function at 810 nm with FWHM of 30 nm, filter D.

3.2. Test

To implement and test a de-convolution algorithm for target spectrum reconstruction which can be considered as an initial part of a validation process, we took the measurements of continuous light source in comparison with measurements taken with same conditions with reference spectrometer. Using our numerical algorithm (software), we could get preliminary results of the reference source spectrum, we initially measured using our prototype instrument and in comparison we used traditional spectrometer (Instrument Systems, model: CAS-140-CT) measurements, both of the measurement results shown in figure 9.

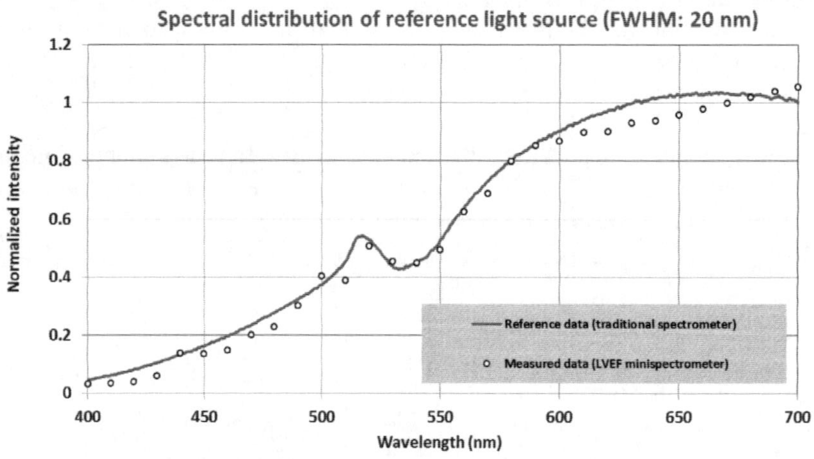

Figure 9. Spectrum of reference light source from 400 nm to 700 nm measured with our compact slit-less mini-spectrometer in comparison with measurements of traditional spectrometer.

4. Conclusions

Prototype instrument of a compact, slit-less mini-spectrometer is demonstrated. It comprises LVEF and CCD array. Characterization of the LVEFs using monochromatic light proves the operation of the LVEF reported. We determined the spectral transmittance and slit functions to realize our concept. This new spectrometer is expected to have a potential of modern applications.

Placing the LVEF directly on the top of a detector array strongly reduces the problems associated with diverging light. Thin film deposition over large format sensor surface can be a good alternative which can be a part of future work. Also the radiometric calibration to obtain a calibration matrix assigning absolute irradiance values to each ROI to get the true spectrum of the test samples is subject of future work.

References

[1] A S.Jakobs, U.Shallenberg, “High precision longpass filter arrays for miniature spectrometers in “Advances in Optical Thin Films” SPIE Proc. 5250, 663-669 (2003).
[2] A.Piegari, J.Bulir, Thin film filters with variable transmittance for spectrometry, IEEE Proc. “Workshop FIBRES and OPTICAL PASSIVE COMPONENTS” (Palermo, Italy, 22-24 June 2005), 109-114 (2005).
[3] J.P. Coates, “New Microspectrometers,” Spectroscopy 15, 21-27 (2000).
[4] J.H. Correia, G. de Graaf, S.H. Kong, M. Bartek and R.F. Wolffenbuttel, “Single-chip CMOS optical minispectrometer”, Sensors and Actuators 82 (2000) 191-197.
[5] R F Wolffenbuttel, “MEMS-based optical mini- and microspectrometers for the visible and infrared spectral range”, J. Micromech. Microeng. 15 (2005) S145-S152.
[6] A. Emadi, H. Wu, G. de Graaf and R.F. Wolffenbuttel, “IR Microspectrometers based on Linear-Variable Optical Filters”, Procedia Engineerig 25 (2011) 1401-1404.
[7] A.Piegari, J.Bulir, “Variable narrow-band transmission filters with wide rejection band for spectrometry”, Appl. Opt., 45, 16 (2006)
[8] A. Emadi, H. Wu, S. Grabarnik, G. de Graaf, K. Hedsten, P. Enoksson, J.H. Correia and R.F. Wolffenbuttel, “Fabrication and characterization of IC-Compatible Linear Variable Optical Filters with application in a micro-spectrometer.”
[9] K. Mahmoud, S-C Park, D-H Lee, “Design of a new compact spectrometer based on a linearly variable edge filter”, Proceedings of the 13th International Conference on New Developments and Applications in Optical Radiometry (NEWRAD, 2017) , pp. 222-223.
[10] Roman Z Morawski, “Spectrophotometric applications of digital signal processing,” Meas. Sci Technology. 17 (2006) R117-R144.
[11] Keivan Ansari, “Spectral band selection using genetic algorithm based weiner filter estimation method for reconstruction of Munsell spectral data, Proceedings of International Symposium on Electronic Imaging (2017) 190-193.
[12] Daniel Massicotte, Roman Z. Morawski, Andrzej, “Kalman-Filter-Based algorithms of spectrometric data correction – Part1: An alternative algorithm of deonvolution”, IEEE TRANSACTIONS ON INSTRUMENTATION AND MEASUREMENT, VOL. 46, NO. 3, JUNE 1997.
[13] Markus Hegland and Robert S Anderssen, “Resolution enhancement of spectra using differentiation”, 2005 Inverse Problems 21 915.

A Multi-Purpose, Detector-Based Photometric Calibration System for Luminous Intensity, Illuminance and Luminance

Brenda H.S. Lam [1], Steven S.L. Yang [1] and Y.C. Chau [1]

[1]Standards and Calibration Laboratory, the Government of the Hong Kong Special Administrative Region, Hong Kong

Abstract. A multi-purpose detector based calibration system for luminous intensity, illuminance and luminance has been developed at the Government of the Hong Kong Special Administrative Region, Standards and Calibration Laboratory (SCL). In this paper, the measurement system and methods are described. The measurement models and contributory uncertainties were validated using the Guide to the Expression of Uncertainty in Measurement (GUM) framework and Supplement 1 to the GUM – Propagation of distributions using a Monte Carlo method in accordance with the JCGM 100:2008 and JCGM 101:2008 at the intended precision level.

1. Introduction

In photometry, candela is the base unit of the International System of Units for luminous intensity, which is used to quantify the performance of luminous intensity standard lamps. Illuminance and luminance are the other two fundamental photometric quantities which are commonly used for light assessments of workplaces and characterization of the uniformity of luminous intensity standard lamps respectively. A single detector-based calibration system for the calibration of illuminance meters, luminance meters and luminous intensity of luminous intensity standard lamps has been developed at the SCL. By comparing the developed system with the traditional method using a standard luminous intensity standard lamp as the reference (a source based system), the developed system has the advantage of relatively simple setup for the calibration of luminous intensity standard lamps, illuminance meters and luminance meters.

2. Measurement Setup

The photometric calibrations of luminous intensity and illuminance in the SCL are performed on a long rail system with movable carriages as shown in Fig. 1. The long rail is seated on a rigid surface to minimize the effects due to mechanical vibration. A precision reference photometer head is mounted on a movable carriage which is allowed to move along the railway. The position of the carriage on the rail is monitored by a distance measuring system that provides the absolute position with a resolution of 0.01 mm. The encoder reading is verified by comparison with a laser distance meter, and the expanded measurement uncertainty is determined to be less than 0.2 %.

The reference photometer is thermostatic stabilized to minimize temperature effect. It is calibrated by Physikalisch-Technische Bundesanstalt (PTB) biannually and has an overall uncertainty of 0.9 %. A precision colorimeter is mounted at the end of the rail to monitor the correlated colour temperature of the luminous intensity standard lamp (Osram Wi 41/G or FEL 1000 W) at 2856 ± 50 K. The uncertainty due to the instability of luminous intensity standard lamp is less than 0.1 %. The standard lamp holder allows 5 axis alignments with very fine adjustment capability to simplify the alignment of the standard lamp. Figure 2 shows the photo of the stage for mounting standard lamp in the SCL. The whole system is accommodated inside a dedicated test chamber, and the light shield enclosure is covered with black velvet. Two to three screens with different aperture sizes are mounted on the rail to reduce stray light effect.

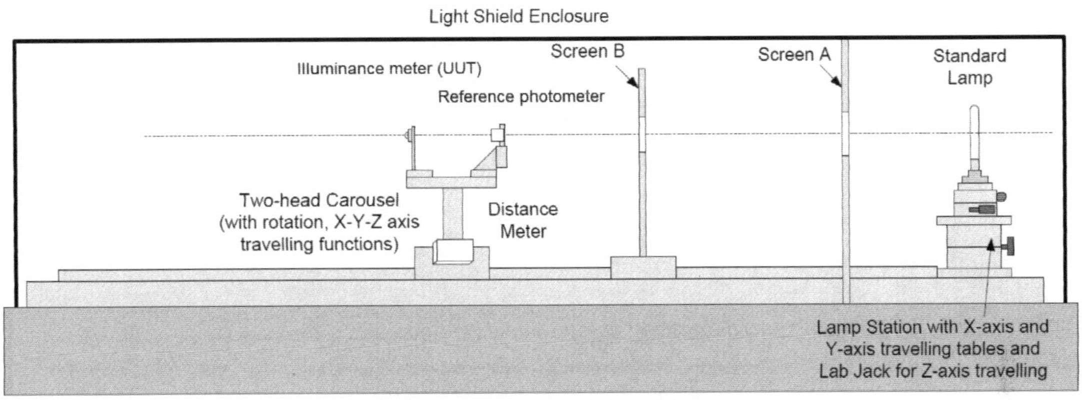

Figure 1. Measurement setup in the SCL

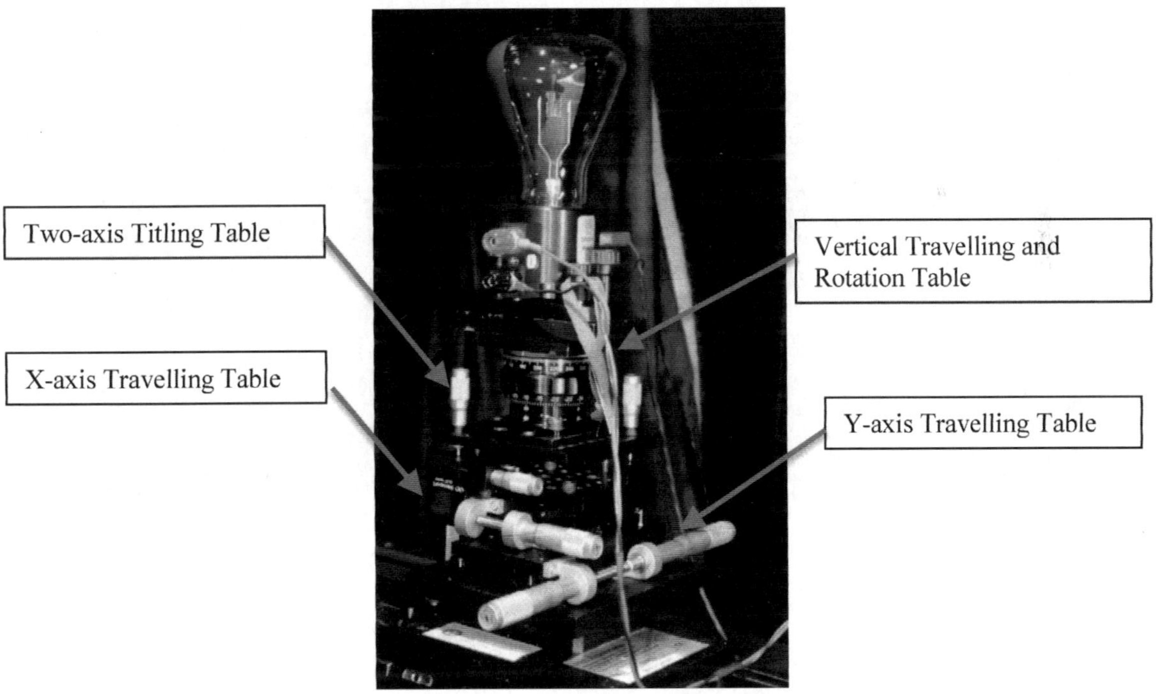

Figure 2. Photo of stage for mounting standard lamp

For the calibration of luminance meter, the calibration system is setup on a 3-meter photometric table as shown in Fig 3 and 4. A 12-inch integrating-sphere operated at 2856 ± 50 K is used as a uniform light source. Four tungsten filament lamps with different power ratings are installed in the sphere to cover the luminance range from 10 cd·m^{-2} to 4000 cd·m^{-2}. Two of them are located separately in two intermediates 3-inch spheres which are attached to the large sphere. Shutters with adjustable irises are located between the small spheres and the large sphere in order to control the output irradiation level of the sphere-source. A precision aperture is placed on the exit port of the large sphere. The reference photometer located in front of the exit port of the sphere is used on the realization of the luminance unit.

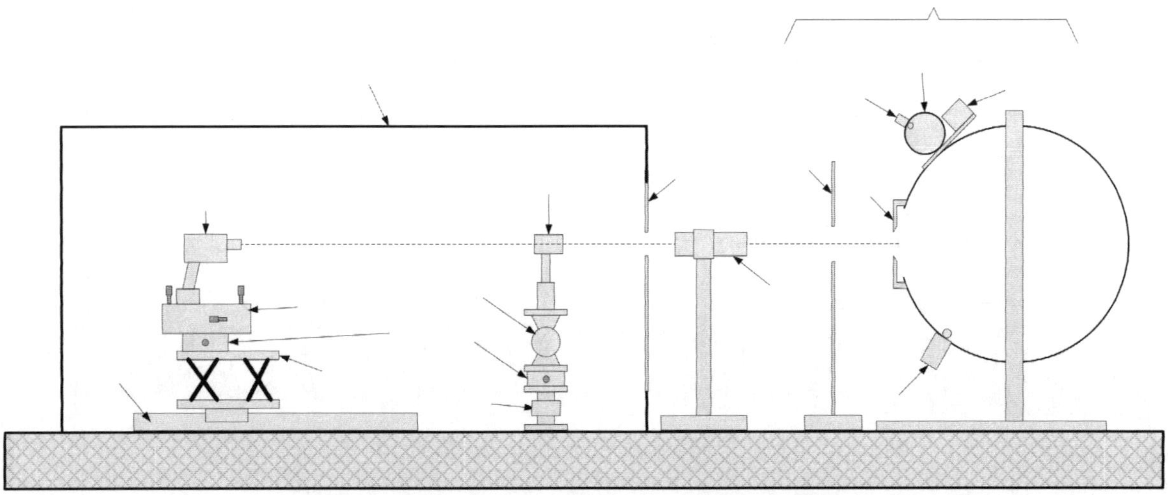

Figure 3. Set-up for realization of the luminance

Figure 4. Photo of the SCL luminance calibration system

3. Measurement Models

The measurement models of luminous intensity, illuminance and luminance calibration are listed in equation (1), (2) and (3) below respectively.

$$I_v = \frac{i}{S}(r)^2 \tag{1}$$

$$E_{X\_C} = \frac{\frac{i}{S} \cdot r^2}{(r + x)^2} - E_{X\_rdg} + C_{stab} \tag{2}$$

$$L_V = \frac{E_V(r_1^2 + r_2^2 + d^2)}{A} \quad (3)$$

where:

I_v is the luminous intensity of the unit under test (UUT),
i is the reading of the reference photometer,
S is the illuminance responsivity of the photometer,
r is the distance between the lamp filament and the front edge of the detector head,
E_{X_C} is the correction for the reading of the UUT (lx),
x is the distance difference between the reference plane defined by the reference photometer head to that of the UUT detector reference plane,
E_{x_rdg} is the UUT reading (lx),
C_{stab} is the correction due to instability of the light source (lx),
L_v is the luminance at the aperture mounted on the exit port of the uniform light source,
E_v is the illuminance at the distance d between the aperture plane of the uniform light source and the photometer head,
r_1 and r_2 are the radius of the diffuser on the photometer head and the aperture of the uniform light source respectively, and
A is the area of the aperture.

4. Evaluation of Measurement Uncertainties

The uncertainty contributions are due to the reference photometer, distance measurement between luminous intensity standard lamp and reference meter, instability or non-uniformity of the luminous intensity standard lamp, alignment error and stay light error are dealt with in accordance with GUM. According to GUM Supplement 1 (GS1), Monte Carlo method (MCM) is alternative to the GUM uncertainty framework (GUF) when the linearization of the measurement model does not provide an adequate representation or the probability density function of the output quantity departs appreciably from a Gaussian distribution or a scaled and shifted t-distribution. A software tool is developed for implementation of MCM to validate the GUF in accordance with the JCGM 101. It is based on MATLAB, which uses the Mersenne Twister algorithm to generate pseudo-random numbers from a rectangular distribution. The algorithm produces a sequence of 32-bit integers with a long period of $2^{19937} - 1$ and passes a comprehensive test for statistical randomness. A graphical user interface (GUI) is developed to facilitate users to generate an MATLAB-Script file for MCM computation. The coverage interval obtained by the MCM is compared with that obtained by GUF. The graphical display of the MCM result is showed in figure 5. We found that for all cases, the GUF is considered valid for the intended use. The major uncertainty components for luminous intensity, illuminance and luminance are listed in Table 1, 2 and 3 respectively. The best measurement uncertainty for the calibration of luminous intensity standard lamps, illuminance meters and luminance meters are estimated as 1.5% with k =2.

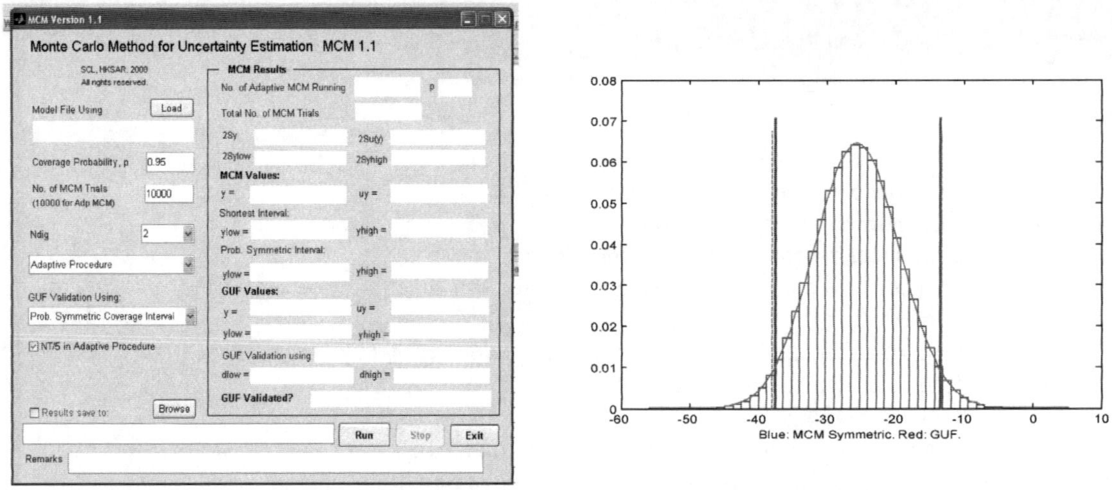

Figure 5. The screen shot of the SCL developed software tool and the graphical display of MCM Result: Using Approximation in Adaptive Monte Carlo Procedure with 10^{-2} Numerical Tolerance for Stability Test

Table 1 Significant uncertainty components for luminous intensity

Quantity	Description	$u(x)$	Distribution
i	Repeatability	0.01 %	t
i	Resolution	0.05 %	rectangle
r	Calibration	0.21 %	rectangle
S	Calibration	0.45 %	normal
S	Drift	0.46 %	rectangle
	Alignment	0.2 %	normal

Table 2 Significant uncertainty components for illuminance

Quantity	Description	$u(x)$	Distribution
i	Repeatability	0.04 %	t
i	Resolution	0.02 %	rectangle
S	Calibration	0.45 %	normal
S	Drift	0.46 %	rectangle
E_{x_rdg}	Repeatability	0.05 %	t
E_{x_rdg}	Resolution	0.03 %	rectangle
C_{stab}	Stability	0.06 %	rectangle
	Alignment	0.36 %	normal

Table 3 Significant uncertainty components for luminance

Quantity	Description	$u(x)$	Distribution
i	Repeatability	0.06 %	t
i	Resolution	0.03 %	rectangle
S	Calibration	0.45 %	normal
S	Drift	0.46 %	rectangle
r_1	Calibration	0.0009 %	normal
r_2	Calibration	0.0003 %	normal
d	Calibration	0.15 %	normal
A	Calibration	0.01 %	normal
A	Temperature	0.012%	normal
	Alignment	0.058 %	rectangle
	Stray light	0.12 %	rectangle
	Uniformity of light source	0.058%	rectangle

5. Conclusions

The single detector-based calibration system is setup at the SCL. The calibration range for luminous intensity, illuminance and luminance are 35 cd to 3500 cd, 1 lx to 3000 lx and 10 cd·m^{-2} to 4000 cd·m^{-2} respectively. The measurements are carefully evaluated in accordance with the JCGM 100 (GUF) and validated the measurements results with JCGM 101 (MCM). The best measurement uncertainty is estimated as 1.5%. This measurement uncertainty and the wide range of measurements are considered adequate for most measurement applications.

References

[1] JCGM 100:2008 and JCGM 101:2008 *Evaluation of measurement data – Guide to the expression of uncertainty in measurement* and *Evaluation of measurement data – Supplement 1 to the "Guide to the expression of uncertainty in measurement" – Propagation of distributions using a Monte Carlo method.*

[2] Ohno Y, Cromer C L, Hardis J E and Eppeldauer G 1994 The detector-based candela scale and related photometric calibration procedures at NIST *Journal of the Illuminating Engineering Society* **23-1** 89-98

[3] Hovila J, Manninen P, Poikonen T and Kärhä P 2015 *Instruction Manual of Luminance and Spectral Radiance Calibrations* (Metrology Research Institute, School of Electrical Engineering, Aalto University)

IOP Conf. Series: Journal of Physics: Conf. Series **972** (2018) 012028 doi:10.1088/1742-6596/972/1/012028

METHOD OF REPRODUCTION OF THE LUMINOUS FLUX OF THE LED LIGHT SOURCES BY A SPHERICAL PHOTOMETER

M. Huriev[1], P. Neyezhmakov[1,2]

[1]NSC "Institute of Metrology", Kharkiv, Ukraine;
[2]KNAME, Kharkiv, Ukraine

mykola.huriev@metrology.kharkov.ua;
pavel.neyezhmakov@metrology.kharkov.ua

Abstract In connection with transition to energy-efficient temporally stable light-emitting diodes (LEDs) lighting, a problem of ensuring the traceability of results of measurement of characteristics of light sources arises. The problem is related to existing measurement standards of luminous flux based on spherical photometers optimized for the reference incandescent lamps with a relative spectral characteristic different from the spectrum of the LEDs. We propose a method for reproduction of the luminous flux, which solves this problem.

1. Introduction

Calibration of photometric lamps in the leading laboratories of the world is performed using spherical photometers [1 - 4]. They are typically represented as a sphere with a diameter of 1.5 meters or more. Inside they are coated with BaSO4 or spectralon having close to 1 diffuse reflectivity in the visible spectrum and scattering phase function, close to equal. As an example, Figures 1 and 2 show the spherical photometer of the national primary standard of luminous flux of Ukraine.

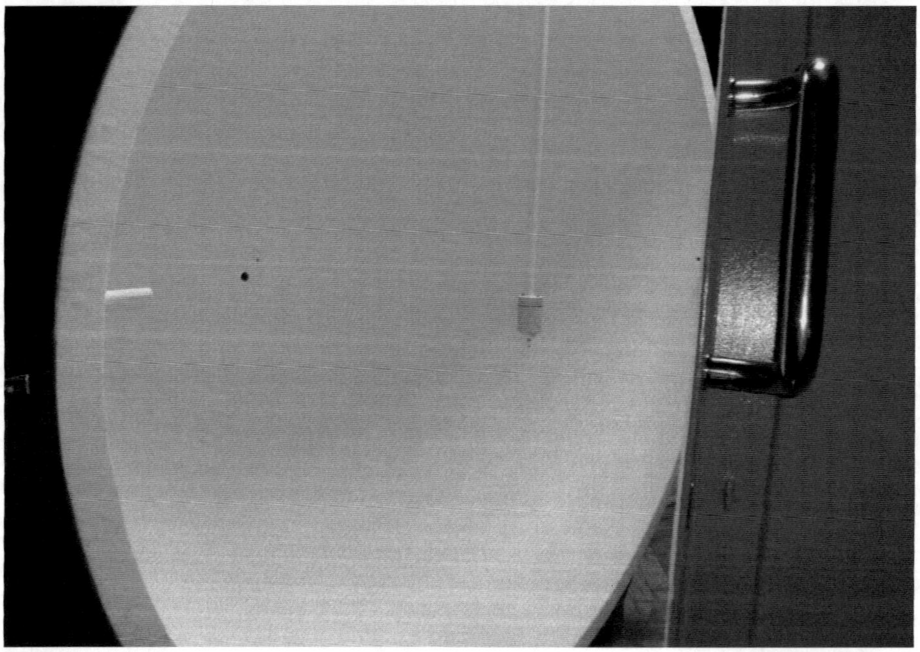

Figure 1. Spherical photometer of the national primary standard of luminous flux of Ukraine

Figure 2. National primary standard of luminous flux of Ukraine

A photodiode with a correction filter and milk glass, which reduces the angular dependence of the sensitivity of the receiver, is fixed to the wall of the sphere. At this, relative spectral sensitivity of the spherical photometer can be represented as:

$$S_{sph.rel}(\lambda) = S(\lambda) \cdot \tau_{mg}(\lambda) \cdot \tau_f(\lambda) \cdot \frac{\rho(\lambda)}{1-\rho(\lambda)} \quad (1)$$

Where $S(\lambda)$ is relative spectral sensitivity of the photodiode, $\tau_{Mg}(\lambda)$ is spectral transmittance of the milk glass, $\tau_f(\lambda)$ is spectral transmittance of the correction filter, $\rho(\lambda)$ is spectral diffuse reflectance of the wall of the sphere.

Such design allows to install the lamps to be measured inside the sphere and to measure its full luminous flux, provided prior calibration of the spherical photometer by a known directional luminous flux.

Figure 3 shows a diagram of reproducing the unit of luminous flux.

For the calibration of spherical photometers, either reference lamps with known luminous flux (type A source) or measured pre-focused light flux from type A source are used. Then calibrated photometers measure the filament lamps with spectral characteristics close to type A source [1 - 4]. The actinic coefficient of spherical photometer to directional light flux is determined by the expression:

$$A = \frac{\int_{380}^{780} \Phi(\lambda) \cdot V(\lambda) d\lambda}{\int_0^{\infty} \Phi(\lambda) \cdot S_{sph.rel}(\lambda) d\lambda} \quad (2)$$

Where $\Phi(\lambda)$ is relative spectral characteristic of the directional luminous flux.

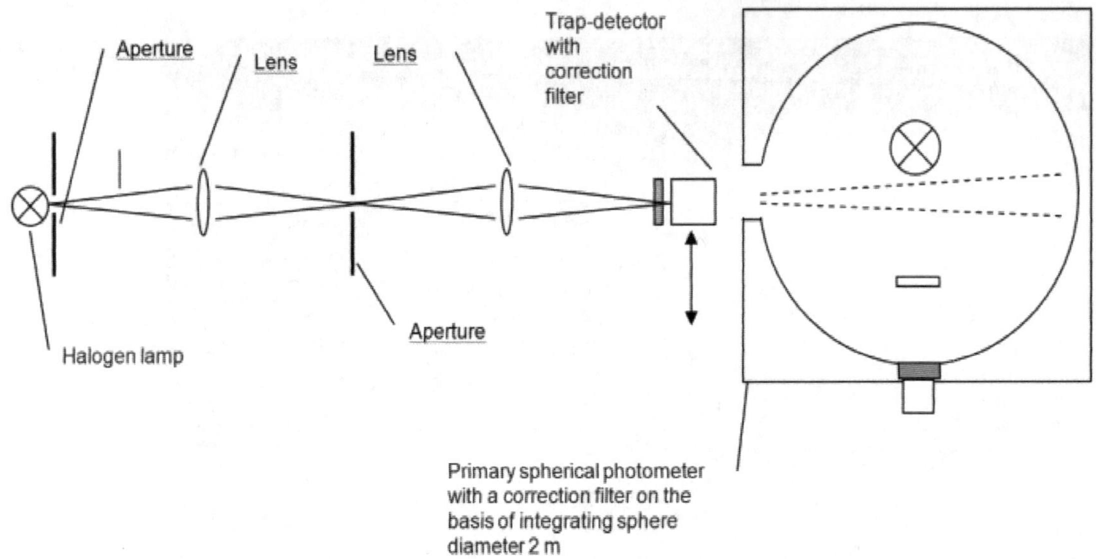

Figure 3. Functional diagram of apparatus for reproduction and transmission of luminous flux to the light sources

The actinic coefficient of spherical photometer to the light flux of the calibrated lamp is determined by the expression:

$$A_l = \frac{\int_{380}^{780} \Phi(\lambda)_l \cdot V(\lambda) d\lambda}{\int_0^{\infty} \Phi(\lambda)_l \cdot S_{sph.rel}(\lambda) d\lambda} \quad (3)$$

Where $\Phi(\lambda)_l$ is relative spectral characteristic of the luminous flux of the lamp to be calibrated.
There will be no significant uncertainty of measurements even with the considerable difference $S_{sph.rel}(\lambda)$ from $V(\lambda)$ in the calibration of incandescent lamps. As $\Phi(\lambda)$ and $\Phi(\lambda)_l$ will be approximately the same for the incandescent lamps, the actinic coefficients of spherical photometer for directional light flux A and light flux of lamp A_l to be calibrated will approximately be the same.

But during calibration of the LEDs there can be significant problems due to the difference of their spectrum from that of incandescent lamps.

Even a slight change of $\rho(\lambda)$ with time leads to the considerable change of $S_{sph.rel}(\lambda)$ (this can be seen from the formula (1)), which begins to affect more its difference from $V(\lambda)$, and therefore the difference between actinic coefficients A and A_l increases.

This is confirmed by interlaboratory comparison (IC 2013) conducted by the International Energy Agency (IEA) with the participation of more than one hundred laboratories, which measured the LED sources and found that differences in the measured luminous fluxes were within ± 5%.

2. Method of realisation and transfer of the luminous flux unit to LEDs

Unlike a spherical photometer, the spectral responsivity of the trap-detector [5] with correction filter (Fig. 4 and 5) is well known, close to $V(\lambda)$, and can be investigated on the spectrophotometer radiometric standard. So it can be used to measure the directional luminous flux from the LED source with the accuracy not worse than that from an incandescent lamp.

It is proposed to use a LED lamp that is similar (by relative spectrum of radiation) to that to be measured as a source of directional luminous flux in the circuit in Figure 2 or to use the same lamp in both cases (Fig. 5).

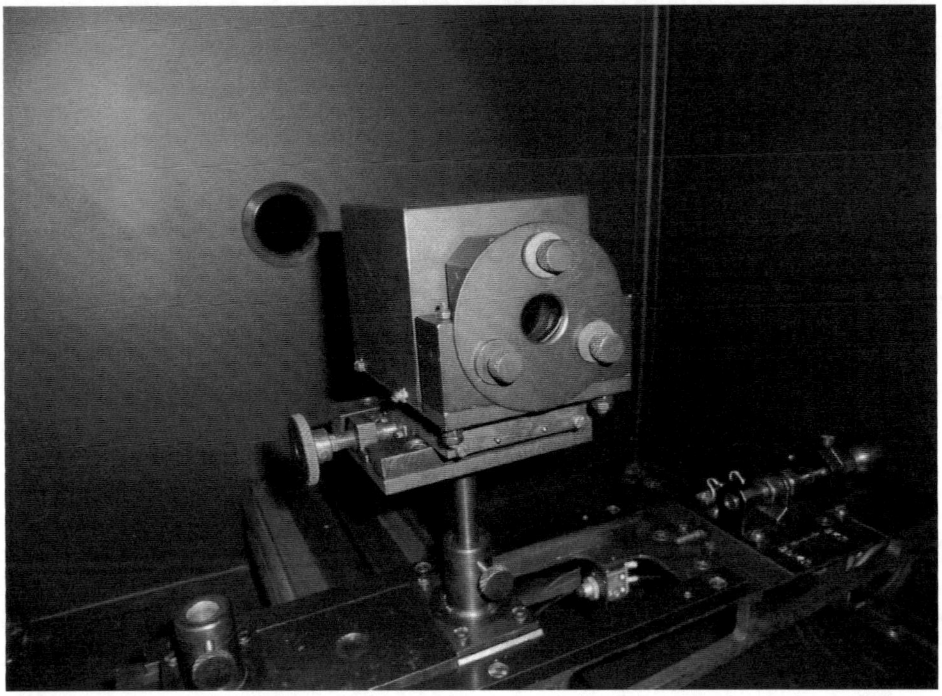

Figure 4. Trap-detector with correction filter

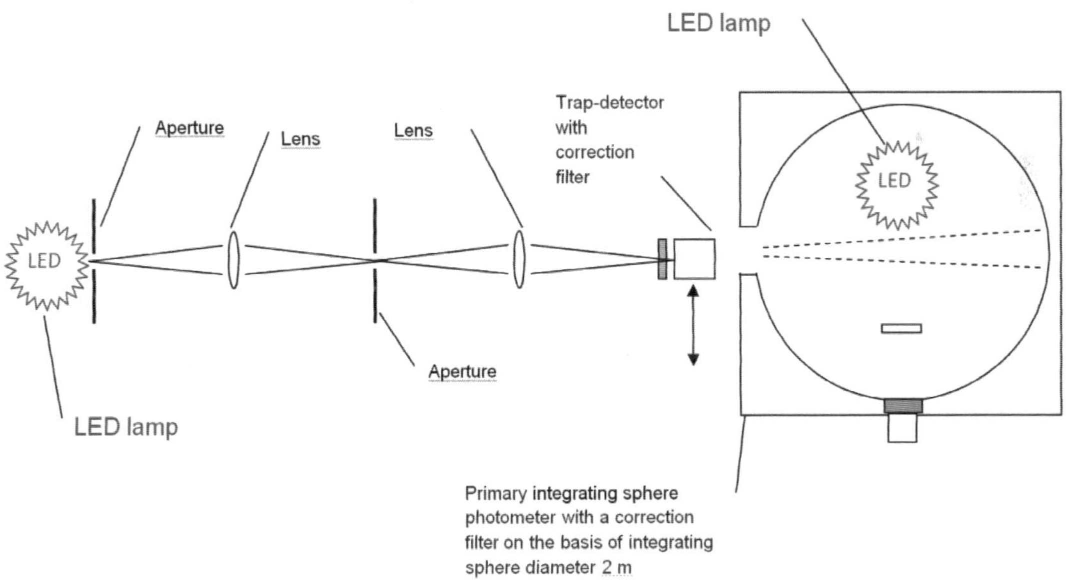

Figure 5. Functional diagram of apparatus for reproduction and transmission of luminous flux to the LED sources

This allows to achieve the same actinic coefficient of spherical photometer for directional luminous flux and the LED lamp to be measured and will provide the necessary accuracy of measurement of luminous flux.

The creation of reference LED lamps is a very difficult task in view of the lack of long-term stability of the luminous flux of these lamps. In this connection, the following design of the reference lamps with high-stable incandescent lamps shown in Figure 6 is proposed for reproduction, transfer, and maintenance of the units of the measurement standards of luminous flux for LED sources. For the reference source of directional luminous flux, we propose to use a correction filter that provides the same spectral characteristic as that of a LED lamp (Fig. 6 (1)).

For the reference lamp that maintains the unit of luminous flux, we propose to use filters that adjust the spectral characteristics of luminous flux to the spectral characteristics of standard LED lamp (Fig. 6 (2)).

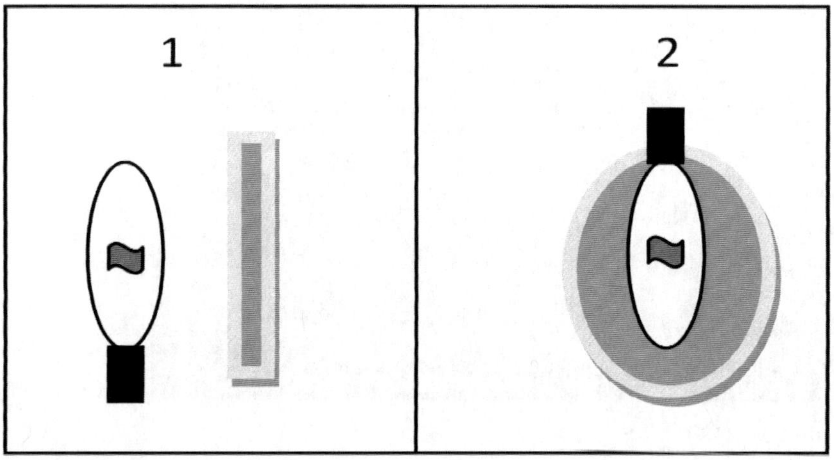

Figure 6. Design of the reference lamps: 1 – for directional light flux; 2 – for measurement inside of the spherical photometer

In this case, the functional diagram of reproduction and transmission of units of luminous flux for LED sources of radiation based on standard filament lamps with corrective filters is shown in Fig. 7.

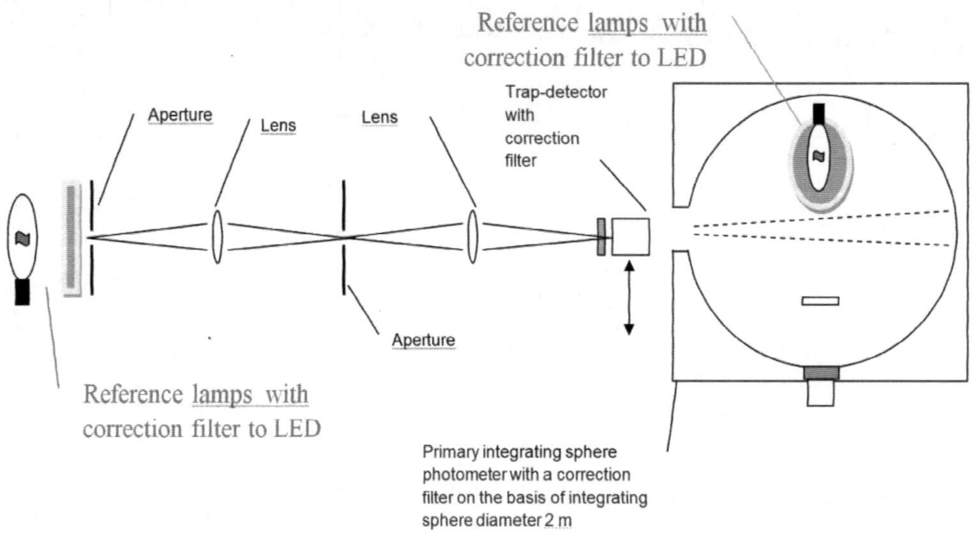

Figure 7. Functional diagram of apparatus for reproduction and transmission of luminous flux to the LED sources based on standard filament lamps with corrective filters

IOP Conf. Series: Journal of Physics: Conf. Series **972** (2018) 012028 doi:10.1088/1742-6596/972/1/012028

3. Conclusion

The proposed method of reproduction and transfer of the luminous flux will allow to reduce the uncertainty of measurements of luminous flux of LED sources to 1%.

References

[1] Yoshihiro Ohno. Nist Measurement Services: Photometric Calibrations. NIST Special Publication 250-37.
[2] Ohno Y. "New Method for Realizing a Total Luminous Flux Scale using an Integrating Sphere with an External Source," *J. IES*, 24-1, 106-115 (1995).
[3] Ohno, Y., "Realization of NIST Luminous Flux Scale Using an Integrating Sphere with an External Source," *the CIE Proceedings, 23rd Session, New Delhi* (1995).
[4] Ohno, Y., "Realization of NIST 1995 Luminous Flux Scale using Integrating Sphere Method," *J. IES*, 25-1, 13-22 (1996).
[5] M. Huriev, A. Kupko, and L. Nazarenko. Double-diode configuration of self-calibrating photodiodes // SPIE Annual Meeting. Advanced Characterization Techniques for Optics, Semiconductors, and Nanotechnologies. Volume 5188.-San Diego, CA, USA-2003. - P. 351-357
[6] Zalewski E.F. Silicon photodiode absolute spectral response self-calibration/ E.F. Zalewski, J. Geist//Appl. Opt., 1980, 19, pp. 1214 – 1216.
[7] Geist J. Spectral response self-calibration and interpolation of silicon photodiodes/ J. Geist, E.F. Zalewski, A.R. Schaefer//Appl. Opt., 1980, 19, pp. 3795 – 3799.